914.204
Gil

145250

Gilbert, Edmund
William

British pioneers
in geography

DATE DUE

*BRITISH PIONEERS
IN GEOGRAPHY*

British Pioneers in Geography

EDMUND W. GILBERT

Emeritus Professor of Geography, University of Oxford
Emeritus Fellow of Hertford College

Barnes and Noble, Publishers New York
Founded *1873*

ISBN 389 04618 3

Set in 11/13pt Garamond
and printed in Great Britain

First Published in the United States, 1972
by Barnes & Noble, Publishers

145250

For three Barbaras

Contents

7

Contents

List of Illustrations

FIGURES AND MAPS

List of Illustrations

Preface

IN A TIME of rapid flux such as the geographical profession has been going through in the last decade or two, there is a real need for people to pause, take stock, restore contact with their roots and reflect on the essential continuity of universal thought over the ages. In Britain (and perhaps America too) this is particularly necessary, not only because the onset of the so-called 'quantitative revolution' has been so precipitate but also because there seems to be less clear agreement on what it is that the national traditions in geography have bequeathed, compared with those of France, Germany or Russia. This lack of a coherent national image may account for the patently inadequate recognition given to the British in general treatises on the history of geographical thought.

This collection of essays by Professor Gilbert will go a long way towards filling this need because although he makes no claim whatever to give a comprehensive account of 'the British contribution', he happens to be concerned with some of the most seminal thinkers which British geography has produced. Most of them are associated with the oldest and one of the most distinctive Schools in the country—that of Oxford, with which Professor Gilbert himself has been associated for half a century. With their traditions of individualism and pragmatism, the idea of a pervasive 'School' of thought deriving from one institution has been less common in Britain and America than in Continental Europe.

Having received my geographical baptism in the Oxford School as an undergraduate (and pupil of Professor Gilbert) I welcome this opportunity to do homage to it and to him. He is a most infectious and sensitive exponent of an approach which has now become unfashionable in some quarters but is more than ever needed as a leavening for contemporary geography. His eloquence as a lecturer was brought home to me in particular when he presented 'Seven Lamps of Geography' (Chapters Seven and Nine in this volume) at the London School of Economics in 1950. The large audience became infected with an almost revivalist fervour for geographical teaching, the like of which is not easily evoked in the academic world these days.

From my present vantagepoint it strikes me that there is an unrecognised but nonetheless real affinity between the Oxford School and perhaps the most distinctive School in American geography over the last half century, that of Berkeley, originating with a deeply humane scholar, Carl Sauer, whom Gilbert quotes more than once in his introduction. When all the obvious differences in such matters as field locale, depth of time-span involved, and so on, are noted, there remains a pervading common concern with creating a philosophy of man as an inhabitant and transformer of the earth. Mackinder's view of geography, scientifically grounded though he was, as 'a philosophy, an art and a literature . . . ranging values alongside measured facts' or Roxby's 'great quality of sympathetic understanding', strike a not unfamiliar chord with the 'Berkeley School'. Techniques and methodologies necessarily change, one hopes for the better, but such continuity of purpose and philosophy remains 'in the walls' of those institutions which have been touched by great personalities.

Gilbert's concentration on notable personalities, their ideas and their context, informs the greater part of this collection, reflecting a *genre* which he has made peculiarly his own in the last two or three decades. His emphasis on good writing, in the common tongue, and rigorously avoiding needless jargon, should be taken to heart by all geographers today. His insistence that geographers learn from regional novelists and gifted amateurs from many other walks of life, while caring for aesthetics and preservation of our environment, also adds to the impression of a broad humanity suffusing this book. We are, and should be, reminded that wisdom and understanding are pearls of greater price than mere 'explanation' or precision in a more mechanical sense, though there is no reason why they should necessarily be antithetical or irreconcilable.

This is in no sense a 'collected works' of Edmund Gilbert, since his major long-term contributions to historical geography, exploration, the geography of cities,* regional planning, and other topics, are unrepresented here. But these essays, on what might be called the theme of geographical biography, or the role of personalities in the making of British geography, make delightful reading and in reality come close to the heart of the matter of how 'disciplines' are really created and nourished.

Berkeley, California   David Hooson
June 1971

---

* Recognised in *Urbanization and its problems: essays in honour of E. W. Gilbert,* ed J. M. Houston and R. P. Beckinsale. Blackwell, Oxford 1968.

Introduction

*'Our reach exceeds our grasp but we do not therefore
reduce our reach'*

CARL O. SAUER, 1971.

THEME OF THE BOOK

INTEREST IN THE history of geography has increased rapidly in
recent years. A Commission of the International Geographical
Union was established at Delhi in 1968 to investigate the
'History of Geographical Thought', and will report at the Congress
to be held in Montreal in 1972. This book has, therefore, a certain
topical interest because it consists of a selection of essays on the
history of geography. Most of the following chapters were originally delivered as lectures and some were given to celebrate anniver-
saries and other special occasions. I have decided to leave the
majority of the essays in almost the same form as that in which
they were delivered. This means that the personal pronoun is
used rather frequently but no apology is made for its retention. In
spite of the fact that the lectures were given at different dates, a
certain unity of theme can be claimed for the book, as each essay
is concerned with the life, work, and writings of one person or a
group of people. Individual geographers must collect their
material in both field and library before they can make their maps,
write their papers and books and draw their conclusions. Both
historical and geographical ideas depend so much on the personality of the man who conceives them.[1] No two persons look
at the same landscape in exactly the same way. The individual
geographer creates the art of geography and it is essential to know

about him, his life and background, if we are to understand his work. In January 1967 I received a friendly letter from Dr D. R. Stoddart of Cambridge in which he said:

> I wish that there were more studies of people rather than ideas divorced from people, since so much to do with the spread of or opposition to ideas seems to result not from their intrinsic worth but from the personalities involved.[2]

He went on to deplore the fact that in the standard histories of geographical thinking, people are absent. Writing in the same vein, Professor Carl Sauer has rightly argued that

> a good knowledge of the work of one or more of our major personalities is about as important an induction into geography as I am able to suggest.[3]

David Stoddart's letter encouraged me to turn a number of my published essays about people into a book. I have been careful not to use the definite article in the book's title. My sole concern here is with British geographers, but it must be emphasised that only a small selection, and that a very personal selection, of British workers in the many broad fields of geography has been made. Some of those about whom I write might not even be regarded as geographers by many of the severely professional geographers in the British universities of today.[4] This book does not pretend to be a comprehensive history of British geographical thought and action. Nevertheless, it must be stressed that in some well-known books by American and other foreign authors on the makers of geographical thought, British geographers are rarely mentioned. Even Hakluyt and Mackinder are sometimes completely ignored. The limelight is usually directed entirely on the German and French schools of geographical thought with an occasional beam on the American. It has to be admitted that the great Russian masters of the subject are often quite as neglected as the British. The purpose of this book is to fill a very small part of this immense gap in our knowledge about British geographers.

What is meant by the word 'pioneer'? In the early sixteenth century a 'pioneer' was one member of a group of foot-soldiers who proceeded ahead of the main army. They carried spades and pickaxes to clear and open a way for the large numbers who were to follow. The *Shorter OED* defines a pioneer as 'one who begins some enterprise, or course of action'; and as 'an original investigator, explorer or worker'; and finally as 'an initiator'. The geographical pioneers discussed in this book were original investigators or initiators; room has not been found for navigators and explorers, and the author regrets that he had not the space for an essay on one of his greatest heroes, Captain James Cook. Britain has produced an immense number of pioneer explorers of both sea and land during the last five hundred years. It is not the object of this book to write about them, as their activities are fully described in many other works.[5] Again there are here no accounts of Christopher Saxton, John Norden, William Smith, John Speed and other famous British pioneers in cartography, although many of them were fascinating characters. The author is especially disappointed that he had not the space to devote an essay to John Ruskin, whose influence on the theory of geographical perception could be compared with Darwin's on geography as a science. Sir Kenneth Clark said of Ruskin 'that in his mind . . . everything was more or less reflected in everything else'.[6] This attitude has been that of many British geographers. J. K. Wright has emphasised both the high development of Ruskin's geographical sense and more especially his 'analyses of the influence of geographic features—particularly mountains—on artistic expression'.[7] 'Landscape painting', said Ruskin in 1871, 'is thoughtful and passionate representation of the physical conditions appointed for human existence. It indicates the aspects and records the phenomena, of the visible things which are dangerous or beneficial to men.' Ruskin 'always saw things both as an artist and a scientist'. This power of double vision is essential for geographers if they are to understand the environment.

The pioneers in this book are all pioneers in geography and it is necessary to define this branch of learning a little more closely. Geography describes and interprets regions and places, both in the past and at the present. The historical element in geography is scientific and objective; it is essential to use history to supplement the subjective and therefore difficult element of description, which brings alive the whole genius of place and region. Work in

Geography should start with detailed observation and experience of life and nature in a clearly defined area. In a letter to the author Carl Sauer complained that in America

> so few geographers begin with observation or get around to it. They are bogged down in numbers which for the most part are not numbers they have collected and so may be the victims of categories they have not established. What you say about the perception of English novelists would fill them with horror as unscientific. We need a different breed who will see geography as one of the liberal arts that is based on attentive seeing and reading.[8]

In another letter to me Sauer complained that in America

> we have raised a brood of quantifiers, model-builders and perceptionists who have never read a page of Ratzel or Cornish.[9]

The rise of the perceptionists, and the vogue for the word 'perception' in the sixties of this century are good reasons for rejoicing, but many of these modern perceptionists do not seem to be aware that geographers have been concerned with perception for a very long time. The quantifiers and model-builders on the other hand often adopt a strangely uncivilised attitude to the history of geographical ideas. In a document that was widely circulated in English schools recently it was stated that 'it is desirable to quantify our statements in order to establish geography as a respectable academic discipline.' One of the objects of this book is to combat the apparently ignorant and arrogant attitude displayed in that statement. Geography has been studied in the ancient English universities for nearly 800 years, and the great tradition of geographical thought goes back to classical antiquity, to Homer, Herodotus and Strabo. Work in geography in Britain has been eminently practical. Until comparatively recently it has been carried out by 'amateurs', by priests and ministers of religion, by British soldiers[10] and administrators in foreign lands, and by gifted men with the leisure provided by private means like Banks, Ford, Galton, Murchison, and Vaughan Cornish. The work has often been applied, to such problems as those of the incidence of disease, the preservation of scenic beauty and, more recently, the detailed investigation of land use. At the same time a sense of poetry or what John Ruskin called 'the instinctive love of nature' has constantly stirred the imagination of the British people to study, not only their own lovely islands, but also the lands and seas of the whole 'vaste globe of the earth'.

CHAPTER ANALYSIS

A brief summary of the contents of this book may help the reader. None of the persons discussed below, not even Mackinder or Roxby, possessed academic qualifications in geography as such. British Honours degrees in geography were only introduced as recently as 1917. This book does not include accounts of the work of the living or of the recently dead. Further it unashamedly concentrates its main attention on the history of geography at Oxford; as already stated, it is not a comprehensive account of British geographical thought and it does not pretend to be complete even for Oxford. The first chapter attempts to put Richard Hakluyt's work in its Elizabethan setting, and to show that the study of geography in his time had immense political significance. The chapter is introduced by accounts of the first lectures on geography given at Oxford by Gerald of Wales in the twelfth century; and also of the work of William Merle of Merton College, the first man in the world to keep a regular daily journal of the weather. He did this for seven years between 1337 and 1344. For example, he wrote a summary of the weather during the last three months of the year 1342:

> There was spring-like weather the whole time between September and the end of December, except on those days to which frost is ascribed, so much so that in certain places leeks burst forth into seed, and in others cabbages blossomed.[11]

Without simultaneous daily observations of the weather in many places the geography of climate cannot prosper.

All the persons described in the first chapter were ordained priests, as were many of those whose work is discussed in Chapter Two under the title 'Geographie is better than Divinity'. A number of the latter were members of either Hart Hall or Magdalen Hall, Oxford. But the list contains several who were not ordained, including Robert Plot, the British pioneer of regional geography, whose writings were based on both observations made on horseback in the field, and answers to his questionnaires. Similarly John Caswell of Hart Hall was the first in Britain to measure accurately the heights of mountains—a pioneer quantifier.

The third chapter turns to one example of the achievements of practical geographers. British administrators of colonial territories did much to alter the human geography of lands under their control; it is sometimes forgotten today that their activities were often of great benefit to the native inhabitants. Richard Kane, Governor of Menorca, was a pioneer in the art of benevolent colonial administration. He built roads, moved the capital to a more convenient site and held a census. Similar work to Kane's was repeated in many other British possessions. For instance the British Protectorate of the Ionian Islands (1815–1864) left a similar legacy of roads, bridges, palaces, country houses and a fine aqueduct built by Sir Frederick Adam, the second High Commissioner between 1824 and 1832. One of the last High Commissioners of the Ionian Islands in 1858–9 was William E. Gladstone, a diligent observer of the environment and geographical processes. As a young man he wrote a graphic account of an ascent of Etna at the commencement of the eruption of 1838.[12] The fourth chapter on medical geography deals with Victorian pioneers in a branch of geography which has been rapidly developed in our own day by the Royal Geographical Society. Hakluyt had seen the need for research in tropical medicine; Captain Cook's practical efforts to preserve health at sea secured him the award of the Copley Medal from the Royal Society. The Victorian doctors brought cartographic methods to the aid of medicine; several including H. W. Acland were Fellows of the Royal Geographical Society. Similarly from 1958, when a committee on Medical Geography was established under the aegis of the Royal Geographical Society, the main task was to produce a *National Atlas of Disease Mortality*. This is undoubtedly one of the two great British contributions to geographical scholarship since the war, the Hakluyt Society's four volumes of Captain Cook's Journals being the other.[13]

The next two chapters are concerned with different aspects of literary geography.[14] Richard Ford's *Hand-book* on Spain has recently been described by W. G. Hoskins as 'the finest travel-book in English': it certainly evokes the spirit of the regions of Spain. Sir William Stirling-Maxwell classed Ford's *Hand-book* as 'among the best books of travel, humour and history, social, literary, political and artistic, in the English language'.[15] It was recently stated that Ford's book 'keeps its place on the reading list recommended to members of Her Majesty's Diplomatic Service posted to Spain; besides informing them about the country it may be hoped that their dispatches will gain from imitating Ford's lucid and racy style'.[16] Ford's skill as an artist is illustrated by the Plates on page 105 and by Fig 14. It is worth noticing that Mackinder always insisted that a geographer should have 'an artistic appreciation of land forms, obtained, most probably, by pencil study in the field'.[17] The standard of art achieved by both nineteenth-century geologists, like Archibald Geikie, and doctors like H. W. Acland, was high. In these days of the universal camera, geographers would do well to remember Ruskin's encouragement to the amateur artist: 'I believe that sight is a more important thing than the drawing; and I would rather teach drawing that my pupils may learn to love Nature, than teach the looking at Nature that they may learn to draw'.[18]

At Auchendavy on the Antonine Wall in Scotland there was found a Roman altar, inscribed by Marcus Cocceius Firmus, a centurion, with the words: GENIO TERRAE BRITANNICAE, that is 'To the Spirit of the Land of Britain'.[19] British regional novelists have often been more successful in evoking the spirit— the geographical individuality—of the diverse regions of Britain than the professional geographers of today. Carl Sauer, in commenting on my chapter on regional novels, told me that he enjoyed reading English crime stories, 'not for the plot but because they are likely to give authentic landscapes of countryside and city, rarely found in American mystery writing'. It would be a fair criticism to say that this book has neglected the work of

British poets in interpreting the genius of the landscape. William Wordworth powerfully depicted the scenery of a distinctive part of England, the Lake District. He also wrote a prose description of the physical nature and human geography of that region in a guide-book that is deeply poetic in feeling.[20] Wordsworth was a Romantic poet of Nature. Perhaps George Crabbe is more truly a geographical poet; Byron described him as 'Nature's sternest painter, yet the best'. Crabbe's remarkably accurate descriptions of the landscape of the Suffolk coast should always be read alongside the scientific writings of the Cambridge school of coastal geographers[21] if one is to appreciate the haunting beauty of that flat country and its great expanse of sky. Matthew Arnold, in two poems, immortalised the Cumnor Hills on which he wandered. Brian Harley has discovered that Arnold walked with an early hachured edition of the one-inch OS map in his hand. Arnold himself said 'a good map becomes almost a picture'.[22] If a reference to a living poet may be allowed, it has been truly said that 'what inspired John Betjeman was a sense of place'[23] and that 'the landscape that most appeals to him is the inhabited landscape'.[24] Like a good human geographer 'he cannot see a place without seeing also the life that is lived in it'.[25] A. J. Herbertson had exactly the same philosophy.

The next five chapters deal with British geographical teaching in the century between 1837 and 1945. A chapter on the inadequacies of early Victorian teaching is followed by two chapters on Mackinder's life and work. C. E. Montague summed up Mackinder's achievements in his masterly review of Mackinder's *Britain and the British Seas* (1902) entitled 'the raising of geography from the dead'.[26] Mackinder's views on the heartland and world strategy still echo round the world. A chapter on A. J. Herbertson, the great exponent of regional geography, emphasises his belief that to separate 'the whole into man, and his environment is a murderous act'. Herbertson stoutly asserted that 'there are no men apart from their environment'. This account is followed by a chapter on P. M. Roxby, the pupil of Mackinder and Herbertson. He was above all a teacher and he based his early teaching of regional geography on observations made in the field by bicycle over large parts of East Anglia. Roxby was certainly the best lecturer that I have ever heard. In the words of Wilfred Smith, Roxby's 'memorial is the Department [of Geography on Merseyside], which he founded, which he nourished and which he inspired.[27]

The last chapter deals with Vaughan Cornish whose stature as a geographer has grown steadily with the passage of time. Carl Sauer has urged that young students should immerse themselves in 'the first-hand study of the individual great figures of our past'.[28] He argued that a study of Humboldt and Ritter, of Ratzel and Hahn, of George Marsh and Vaughan Cornish[29] 'will provide a truly liberal geographical education'. Both Marsh[30] and Cornish began their geographical careers in the physical study of waves. Marsh later turned to the investigation of the effects made by man on the face of nature; while Cornish changed his main interest to the beauty of nature. Marsh is worth reading today if only because he objected to the currently fashionable idea that only the measurable is worthy of serious study. Andrew Goudie[31] has shown that Cornish, in the physical branch of his research, had original ideas of great value to later workers. Again Cornish's[32] pioneer work in political geography, especially his concepts of 'storehouse' and 'crossroads' in the investigation of urban problems, paved the way for the many who now write on conurbations and megapolitan areas. But above all Cornish must be remembered for his writings on the beauty of nature and its conservation. He heard that great explorer, Sir Francis Younghusband, when President of the Royal Geographical Society, declare in 1920 that 'knowledge of the beauty of the Earth was the most important form of geographical knowledge'.[33] In response to this eloquent appeal Vaughan Cornish was instantly converted, like St Paul, and devoted the rest of his life to 'the analytical study of beauty and scenery'. This is a matter of vital significance today. In their wisdom a large body of young British geographers have turned away from what they call 'traditional' geography to quantification and the abstract study of building models. The latter occupation seems closely similar to the work of the medieval writers on legendary islands of the Atlantic.[34] The abstract models do not exist in reality any more than did the island of St Brendan. At a time when the world resounds with talk of the need for conservation and the dangers of pollution, geographers are handing over their obvious duty to take part in these campaigns to the ecologists; they have turned their minds to the arid joys of cost-benefit analysis and now ignore the study of the environment as a whole. In the words of Professor Charles A. Fisher, the assumptions of the quantifiers of the second 'New Geography' now 'threaten to transform the whole of humanity into one vast mass of undifferentiated growth-propelled computer-fodder'.[35] But Vaughan Cornish did not work in vain. In the Roskill *Report of the Commission for the Third London Airport* (1971) was a 'Note of Dissent' by Professor Colin Buchanan. In this he said that 'the belt of open country between London and Birmingham'[36] in which a majority of the Commission had recommended the building of an airport at Cublington, 'still possesses its area of deep quietude'. He added that 'it still has the softest and subtlest landscapes to be found anywhere in the world, landscapes which have inspired generations of English poets and painters'.[37] It was a matter of satisfaction to many 'traditional' geographers to know that the spirit of Ruskin, Younghusband and Vaughan Cornish still lives on; and to learn on 26 April 1971 that the proposal to build an airport at the inland site of Cublington had been rejected by Her Majesty's Government. One 'environmental disaster' had been averted. The defence of the natural beauty of the world's landscapes is the inescapable duty of the geographer.

NOTES

1 The important part played in geographical scholarship by the human qualities of the scholars themselves is fully discussed by Wright, J. K., in his delightful *Human Nature in Geography* (1966).

2 Stoddart, D. R., in a letter to the author dated 23 January 1967. Dr David Stoddart himself made an invaluable contribution to

geographical thought in his stimulating paper on 'Darwin's impact on geography', *Ann Ass Am Geog.* 56 (1966), 683–98.

3   Sauer, Carl O., *Land and Life* (1967), 355.

4   In a recent article a comparison was made between Mackinder and his 'contemporary, the non-geographer George Adam Smith, whose study, *The Historical Geography of the Holy Land,* has rarely been equalled and never excelled, even by trained geographers'. The strange idea that geographers can be divided into 'non-geographers' on the one hand, and 'trained' or professional geographers on the other, began in Britain in 1931. Moss, R. P., *S African Geog J.* 52 (1970), 35.

5   Baker, J. N. L., *A History of Geographical Discovery and Exploration* (2nd ed. 1937) includes invaluable summaries of the work of British pioneer explorers.

6   *Ruskin and His Circle* (Arts Council, 1964); Clark, Kenneth, Introduction, 5.

7   Wright, J. K., op cit in Note 1, 22; Ruskin, John, *Lectures on Landscape. Delivered at Oxford in Lent Term, 1871* (1897), 15.

8   Sauer, Carl O., in a letter to the author, dated 3 March 1965; see also Sauer, Carl O., 'The quality of geography,' *The Californian Geographer* (1970), 5–9.

9   Sauer, Carl O., in a letter to the author dated 28 July 1970.

10 See, for example, the admirable paper by Marshall-Cornwall, General Sir James, 'Three soldier geographers,' *Geogr J,* 131 (1965), 357–65. The trio described were W. M. Leake (1777–1860), E. Sabine (1788–1883) and F. R. Chesney (1789–1872). All three were Fellows of the R.G.S.

11 Symons, G. J., a reproduction and translation of *Consideraciones Temperiei Pro 7 Annis* by Merle, William, 1337–44 (London, 1891).

12 Dennis, George, *A Handbook for Travellers in Sicily* (Murray, 1864), 442–7, includes a long extract from Gladstone's journal of his ascent of Etna in October 1838.

13 The Medical Geography Committee of the R.G.S., on which I had the honour to serve for over ten years, supervised the production of the *National Atlas of Disease Mortality in the United Kingdom* (1st ed 1963; 2nd ed 1970), by Howe, G. Melwyn. See also Stamp, L. Dudley, *Some Aspects of Medical Geography* (1964). For Cook see Beaglehole, J. C. (ed), *The Journals of Captain James Cook* (4 vols, Hakluyt Society, 1955–67).

14 Prince, Hugh C., 'The geographical imagination', *Landscape,* (Santa Fé, New Mexico), 11 (1962), 22–5, is an eloquent essay on this theme.

15 Stirling-Maxwell, W., 'Richard Ford', in *Miscellaneous Essays and Addresses*, vol VI (1891), 105. The two essays in this volume that are devoted to Ford (101–17) originally appeared in *The Times*, 4 September 1858, and *The Press*, 11 September 1858.

16 *Times Literary Supplement*, 4 December 1970, No 3588, p 1413.

17 Mackinder, H. J., *Geog J*, 6 (1895), 376.

18 Ruskin, John, *Elements of Drawing* (1892 ed), XV. See also Hutchings, G. E., *Landscape Drawing* (1960).

19 Wright, R. P., *Roman Inscriptions of Britain*, I (1965), No 2175.

20 Wordsworth, W., *A Guide through the District of the Lakes* (5th ed, Kendal, 1835).

21 See, for example, Steers, J. A., 'The east Anglian coast', *Geogr J*, 69 (1927), 24–48, which I enjoyed hearing in 1926.

22 See Sheet 70 Oxford (1969) of the David & Charles reprints of the First Edition of OS one-inch maps, with notes on Matthew Arnold by Dr J. B. Harley.

23 Betjeman, John, *Selected Poems* (1948), with Preface by John Sparrow, xi.

24 Ibid.

25 Ibid.

26 *Manchester Guardian*, 7 February 1902.

27 Smith, Wilfred, *Geography and the Location of Industry* (1952), 2.

28 Sauer, Carl O., *Land and Life* (1967), 355.

29 Ibid, 355–6.

30 Lowenthal, David, *George Perkins Marsh: versatile Vermonter* (1960); Lowenthal, David, 'George Perkins Marsh on the nature and purpose of Geography', *Geogr J*, 126 (1960), 413–16; Marsh, G. P., *Man and Nature* (ed Lowenthal, David, 1965).

31 Goudie, Andrew, 'Vaughan Cornish', *Trans I.B.G.*, 55 (1972).

32 Cornish, Vaughan, *The Great Capitals: An Historical Geography* (1923).

33 *Geogr J*, 56 (1920), 8.

34 Babcock, William H., *Legendary Islands of the Atlantic: A Study in Medieval Geography* (1922).

35 Fisher, Charles A., 'Whither Regional Geography?', *Geography*, 55 (1970), 388; similarly Professor A. E. Smailes, in a notable Presidential Address to the I.B.G., deplored the present 'neglect of the whole dimension of historical time' in explanations in geography; as well as urging geographers to focus their work upon the 'worthwhile intellectual problems presented by the real world', *Trans IBG*, 53 (1971), 1–14.

36 *Report of the Commission on the Third London Airport* (1971), in *Note of Dissent* (149–60), by Buchanan, Colin, 149, para 1. In a

paper by Baker, J. N. L., and Gilbert, E. W., written over twenty-five years ago, the authors pointed out that the doctrine of an axial belt of industry from London to Liverpool, as then postulated by certain geographers, was unsound. The joint authors of the paper stated that there was 'an area of lighter population density between the London district and the Midlands'. *Geogr J*, 103 (1944), 49–72. It was in this break, in the so-called 'axial belt', still obvious today, that the Roskill Commission proposed that the Cublington airport should be built.

37 Buchanan, Colin, op cit, 150, para 8.

CHAPTER ONE

Richard Hakluyt and His Oxford Predecessors

T HE STUDY OF geography at Oxford has a long and honourable tradition; one that is as old as the University itself.[1] Among the first recorded Oxford lectures of any kind are those given in the year 1187, nearly 800 years ago, by Gerald of Wales, who read aloud his *Topography of Ireland* for three whole days. The lectures were followed by junketing and feasting. Gerald was not exactly a modest man and he described the presentation of his lectures at Oxford in these words:

When in process of time the work was finished and corrected, and not wishing to place the candle which he had lit under a bushel, but to lift it aloft on a candlestick that it might shine forth, he determined to read it before a great audience at Oxford, where, of all places in England, the clergy were most strong and pre-eminent in learning. And since his book was divided into three parts (*distinctiones*) he gave three consecutive days to the reading, a part being read each day. On the first he hospitably entertained the poor of the whole town whom he had gathered together for that purpose; on the morrow he entertained all the doctors of the divers faculties, and those of their scholars who were best known and best spoken of; and on the third day he entertained the remainder of the scholars together with the knights of the town and a number of the citizens. It was a magnificent and costly achievement since thereby the ancient and authentic times of the poets were in some manner revived, nor has the present age seen nor does any past age bear record of the like.[2]

A geographical recitation did in fact also take place very early in the history of Cambridge. In about 1253 Michael of Cornwall, a wandering poet, read a poem in praise of Cornwall and England before the Chancellor of Cambridge together with 'the university of masters'. Gerald's *Topography of Ireland* is not only geographical in character; some of its stories are very tall, strange and almost 'improper' if that Victorian epithet can be used today about anything. A recent translator of the *Topography* said of his own translation that 'some of the expressions used in connection with bestiality have been softened without, however, interference with the obvious meaning'.[3] Professor David Knowles has pointed out that Gerald of Wales had a number of characteristics which give him a link with the world of today. In particular he possessed 'a love, or at least a sense of natural beauty and wild landscape', as well as 'a warm affection for his home and for his native land'.[4] Vaughan Cornish, whose work is discussed in the last chapter of this book, developed the study of natural beauty 700 years or more after Gerald. The *Itinerarium Cambriae* or a journey through Wales, and a *Description of Wales*, are Gerald's best known pieces of writing. The first is an account of his experiences with Archbishop Baldwin when they preached the Crusade in Wales in 1188. The lectures on Ireland given by Gerald of Wales at Oxford with such marathon technique are a reminder that the study of geography in Oxford is coeval with the university itself and its three ancient faculties of Theology, Law, and Medicine. Those few who imagine that geography is an entirely new study should remember the writings of Gerald of Wales (?1146–1220).

Among other early Oxford workers in fields closely allied to geography was William Merle, a Fellow of Merton who died in 1347. It is believed that he was the first man in the world to keep a continuous daily journal of the weather. His observations were recorded for seven years from 1337 to 1344.[5] On this record he based some rules for predicting the weather. He was not the first to do this as classical authors, for example Aristotle and Virgil, had forecast the weather. The daily collection of meteorological information is fundamental for the study of geography: it began at Oxford.

The first person known to have held office as Lecturer in Geography at Oxford was Baldwin Norton, who was appointed at Magdalen College in 1541;[6] it is believed that geography was not officially recognised at Cambridge until seven years later. Towards the close of the sixteenth century the study of geography flourished exceedingly at Oxford, largely because of the activities of Richard Halkuyt, Student of Christ Church. It is worth noticing that three of the greatest Oxford geographers have been Christ Church men, Hakluyt, Mackinder, and Roxby. The name Hakluyt has a foreign ring and perhaps a Dutch sound about it. Although it has been suggested that it is Welsh in origin it is almost certainly English. For generations there had been Hakluyts in Herefordshire, at the manor of Eaton near Leominster. The family had supplied sheriffs and MPs for the service of the county from the reign of Edward II. Richard Hakluyt, the geographer, was born in 1552 in London; he was the second son of another Richard Hakluyt, a member of the Skinners' Company, and Margery his wife. The father died when his son was only four or five years old; in his will he commended his family of four sons and two daughters to the care of his nephew, also named Richard Hakluyt, not long enrolled as a member of the Middle Temple. Soon after, the mother also died and the young lawyer, when only about twenty-five years old, took over the care of his orphaned cousins. The practising lawyer seems to have been a sound business man, expert in the wool trade in his native county of Herefordshire. In 1560 Queen Elizabeth made the Abbey of Westminster a collegiate church and there instituted 'a Dean, twelve Prebendaries, a Schoolmaster, and Usher and Forty Scholars called the Queen's Scholars whereof six or more are preferred every year to the Universities'. The young Richard Hakluyt became one of the forty scholars; then at some time between 1560 and 1570 he was one of the six youths preferred to the University; and in due course of time he became one of the twelve prebendaries of the Abbey.[7]

In the Epistle Dedicatorie to *The Principall Navigations* of 1589 Hakluyt describes a visit which he paid to his cousin's chambers

PETRUS HEYLYN. S.T.P.
Ecclesiæ Collegiatæ Sancti Petri Westmonasteriensis
Canonicus, Martyri & superstiti CAROLIS Patri ac Filio,
Magnæ Britanniæ etc. Monarchis, dum viveret, à Sacris.

London Printed for C. Harper at the Flower de-luce in Fleetstreet.

Plate 1   Peter Heylyn

Plate 2   Robert Plot

in the Middle Temple, probably in 1568. He was then a boy of sixteen and was excited to see a world map and geographical books on the table. He recalled the incident in these words:

I do remember that being a youth, and one of her Majesties scholars at Westminster that fruitfull nurserie, it was my happe to visit the chamber of M. Richard Hakluyt my cosin, a Gentleman of the Middle Temple, well knowen unto you, at a time when I found lying open upon his boord certeine bookes of Cosmographie, with an universall Mappe: he seeing me somewhat curious in the view thereof, began to instruct my ignorance, by shewing me the division of the earth into three parts after the olde account, and then according to the latter, and better distribution, into more: he pointed with his wand to all the knowen Seas, Gulfs, Bayes, Straights, Capes, Rivers, Empires, Kingdomes, Dukedomes and Territories of ech part, with declaration also of their special commodities, and particular wants, which by the benefit of traffike, and entercourse of merchants, are plentifully supplied.

The young boy received this lesson in economic geography from his cousin and then:

From the Mappe he brought me to the Bible, and turning to the 107 Psalme, directed mee to the 23 and 24 verses, where I read, that they which go downe to the sea in ships, and occupy by the great waters, they see the works of the Lord, and his woonders in the deepe, etc. Which words of the Prophet together with my cousins discourse (things of high and rare delight to my yong nature) took in me so deepe an impression, that I constantly resolved, if ever I were preferred to the University, where better time, and more convenient place might be ministred for these studies, I would by Gods assistance prosecute that knowledge and kinde of literature, the doores whereof (after a sort) were so happily opened before me.

From about 1560 to 1570 Hakluyt was a scholar at the newly re-founded Westminster School and in 1570 he was elected as one of the two Westminster Queen's Scholars to Christ Church, Oxford. It took him seven years to work for the degrees of BA and MA. After this he was elected to a Studentship (that is a Fellowship) of his College, which he held until at least 1587. He probably gave it up before 1589 as on the title page of his *Principall Navigations* (1589) he describes himself as 'Student sometime of Christ-church in Oxford'. In the same book he tells of his work in the Oxford years:

> According to which my resolution, when, not long after, I was removed to Christ-church in Oxford, my exercises of duety first performed, I fell to my intended course, and by degrees read over whatsoever printed or written discoveries and voyages I found extant either in the Greeke, Latine, Italian, Spanish, Portugall, French, or English languages, and in my publike lectures was the first, that produced and shewed both the olde imperfectly composed, and the new lately reformed Mappes, Globes, Spheares, and other instruments of this Art for demonstration in the common schooles, to the singular pleasure and generall contentment of my auditory.

Hakluyt must have studied modern languages intensively at Oxford; he was also an omnivorous reader. Mackinder, also a Christ Church man, 300 years later, used to say that he was only the second Reader in Geography at Oxford, as Hakluyt was the first. Mackinder always pleaded that an Oxford chair of Geography, if one were ever established, should be called 'the Hakluyt Professorship';[8] it is unfortunate that this suggestion was not adopted. It is unlikely that this change in nomenclature will ever be made, but if more Oxford chairs in Geography were created it would be possible.

Hakluyt was at Oxford for at least thirteen or fourteen years. How did he support himself in these early years? New light has been thrown on this by Tom Girtin in an article in the *Geographical Journal*.[9] While at Oxford Hakluyt held a scholarship from Westminster School, and in addition a very small supplement from the Nowell educational bequest for needy scholars. He also received some support from City Companies; he received from the Skinners' Company, of which his father had been a member, an exhibition of £2 13s 4d a year. He took his BA in 1574 and MA in 1577. He then made contacts with foreign geographers. At some time in 1577 he met the Flemish geographer Ortelius during the latter's visit to England. In 1580 he was in correspondence with Mercator, another distinguished Flemish geographer. It is clear that from 1578 onwards, after he had taken up the study of divinity, he received a Clothworkers' Exhibition. The first quarterly payment of £16 13s 4d was made in 1578. He preached before the Company in 1580 and again in 1581; perhaps it was then that he became known as Richard Hakluyt, Preacher, to distinguish him from his cousin, Richard Hakluyt, Lawyer. Hakluyt had been ordained Deacon and Priest some time between 1577 and 1580, by Bishop Pierce of Salisbury, a former Dean of Christ Church. He was still tutoring in divinity in 1580; when it is recorded that he taught Mr Gabriel Bowman of Magdalen College. There is no doubt that Hakluyt's connection with the Clothworkers' Company was a great asset to him in making contacts with merchants and other prominent people in London.

In 1583 Hakluyt took up an overseas appointment although he was still referred to as a scholar of Oxford. His own words are as follows:

> In continuance of time, and by reason principally of my insight in this study [geography], I grew familiarly acquainted with the chiefest Captaines at sea, the greatest Merchants, and the best Mariners of our nation: by which meanes having gotten somewhat more than common knowledge, I passed at length the narrow seas into France with sir *Edward Stafford*, her Majesties carefull and discreet Ligier, where during my five yeeres abroad with him in his dangerous and chargeable residencie in her Highnes service, I both heard in speech, and read in books other nations miraculously extolled for their discoveries and notable enterprises by sea, but the English of all others for their sluggish security, and continuall neglect of the like attempts especially in so long and happy a time of peace, either ignominiously reported, or exceedingly condemned.[10]

Hakluyt was goaded by this state of affairs and determined to make good the lack of any account of English voyages; he took up this study eagerly and made it his life's work. During his years in France as the Ambassador's chaplain Hakluyt 'busily gleaned every kind of geographical, mercantile and statistical piece of information that he could discover in the listening-post that was France'.[11] He was constantly going to and fro between England and France with secret messages. He was still paid his pension of £6 13s 4d a quarter by the Clothworkers' Company. Girtin has suggested that Hakluyt was a 'commercial spy', whom the Lord Treasurer Burleigh wished to be paid unobtrusively by a City Company, and that Hakluyt's work was a 'cover' for other activities. It is certain that the pension ended on Lady Day 1587.

Hakluyt himself gives an account of his pride in the achievements of English navigators; his fervent patriotism so typical of his time drove him on to vast labours of scholarship. In his *Epistle Dedicatorie* to Sir Francis Walsingham, dated 17 November 1589, Hakluyt writes:

> To harpe no longer upon this string, and to speake a word of that just commendation which our nation doe indeed deserve: it can not be denied, but as in all former ages, they have bene men full of activity, stirrers abroad, and searchers of the remote parts of the world, so in this most famous and peerlesse governement of her most excellent Majesty, her subjects through the speciall assistance, and blessing of God in searching the most opposite corners and quarters of the world, and to speake plainly, in compassing the vaste globe of the earth more than once, have excelled all the nations and people of the earth. For, which of the kings of this land before her Majesty, had theyr banners ever seene in the Caspian sea? which of them hath ever dealt with the Emperor of Persia, as her Majesty hath done, and obteined for her merchants large and loving privileges? who ever saw before this regiment, an English Ligier in the stately porch of the Grand Signor at Constaninople? who ever found English Consuls and Agents at Tripolis, in Syria, at Aleppo, at Babylon, at Balsara, and which is more, who ever heard of Englishment at Goa before now? what English shippes did heeretofore ever anker in the mighty river of Plate? passe and repasse the

unpassable (in former opinion) straight of Magellan, range along the coast of Chili, Peru, and all the backside of Nova Hispania, further than any Christians ever passed, travers the mighty bredth of the South sea, land upon the Luzones in despight of the enemy, enter into alliance, amity and traffike with the princes of the Moluccaes, and the Isle of Java, double the famous Cape of Bona Speranza, arive at the Isle of Santa Helena, and last of all returne home most richly laden with the commodities of China, as the subjects of this now florishing monarchy have done?[12]

There are British geographers today who argue that it is their duty not only to be scholars upholding certain intellectual values, but also to apply their studies to present troubles and even to predict the future. In such activities they are only following in the pioneer footsteps of Richard Hakluyt. He wrote a *Discourse of Western Planting* (1584) in which he asserted that the founding of a prosperous English colony overseas would remedy unemployment: he thus concerned himself with the politics of his day and argued that geographical learning could benefit the State. He wrote about his proposed new colony in great deail as follows:

But wee for all the Statutes[13] that hitherto can be devised, and the sharpe execution of the same in poonishinge idle and lazye persons for wante of sufficient occasion of honest employmente cannot deliver our common wealthe from multitudes of loyterers and idle vagabondes. Truthe it is that throughe our longe peace and seldome sicknes (twoo singuler blessinges of almightie god) wee are growen more populous than ever heretofore: So that nowe there are of every arte and science so many, that they can hardly lyve one by another, nay rather they are readie to eate upp one another: yea many thousands of idle persons are within this Realme, which havinge no way to be sett on worke be either mutinous and seeke alteration in the state, or at leaste very burdensome to the common wealthe, and often fall to pilferinge and thevinge and other lewdnes, whereby all the prisons of the lande are daily pestred and stuffed full of them, where either they pitifully pyne awaye, or els at lengthe are miserably hanged, even

xxti. at a clappe oute of some one Jayle: whereas if this
voyadge were put in execution, these pety thieves mighte be
condempned for certen yeres in the western partes, especially
in newfounde lande in sawinge and fellinge of tymber for
mastes of shippes and deale boordes, in burning of the firres
and pine trees to make pitche tarr rosen and sope asshes, in
beatinge and wockinge of hempe for cordage . . .[14]

Hakluyt thus prescribed emigration as a cure for over-popula-
tion and unemployment. The Queen in person received Hakluyt's
manuscript essay from his hand. It was not a public document
and was not printed until 1877.[15] The *Discourse of Western Planting*
argued that one nation, the English, would occupy both sides of
the Atlantic. The Queen's comments are not recorded but she
decided that she would not act on its proposals. The document
was not known by the people who settled seventeenth-century
America. In Dr J. A. Williamson's words 'it is worth reading now
as a fascinating essay in planning, wisely rejected'.[16]

Another study, of which Hakluyt was an advocate, far in
advance of his time, was tropical medicine and medical geography.
In his dedication of the third volume of his *principal Navigations*
(1600) he tells that he had decided to omit a pamphlet on *The
curing of hot diseases incident to travellers in long and Southern voyages* by
one George Watson, because it was so defective. But he also says
that Doctor William Gilbert, author of the famous *De Magnete*
(1600) had promised that the College of Physicians would pre-
pare a treatise on diseases in both hot and cold regions.[17]

In 1590 Lady Sheffield presented Hakluyt to the country living
of Wetheringsett in Suffolk which he held until his death in 1616.
He collected other ecclesiastical preferments including canonries
at Bristol and Westminster. In 1589, a year after the defeat of
the Armada, he first published his greatest work, *The Principall
Navigations, Voiages and Discoveries of the English Nation*, a book
of over three-quarters of a million words. He kept this book up to
date and in 1598–1600 appeared his much larger three-volume
work *The principal Navigations, Voiages, Traffiques and Discoveries of
the English Nation*. Hakluyt was an editor rather than an original
writer, but the Oxford historian, J. A. Froude, rightly called *The*

principal Navigations 'the prose epic of the modern English nation'. These were not Hakluyt's only works. In 1582 he had published a small book of sixty leaves, *Divers voyages touching the discoverie of America* and two years later he had written his *Discourse on Western Planting*, already discussed; in 1609 appeared his *Virginia richly Valued*, a translation from Hernando de Soto. He left behind an enormous collection of unpublished material which was incorporated in four large volumes by Samuel Purchas in *Hakluytus Posthumus or Purchas his Pilgrimes*, in 1625.

Hakluyt, in Professor E. G. R. Taylor's words, 'was no figure of Elizabethan romance. Indeed, he must be confessed a mere armchair geographer, for he never travelled farther afield than Paris'.[18] Yet in the collection of his first-hand accounts of voyages and journeys he tells of the hardship he himself suffered: 'What restlesse nights, what painefull dayes, what heat, what cold I have indured; how many long and chargeable journeys I have traveiled; how many famous libraries I have searched into; . . . what expenses I have not spared.'

Very little is known about Hakluyt the man because he tells so little about himself he desired no memorial save his book. He signs himself 'Master of Arts and Student sometime of Christchurch in Oxford'. On the title-page of his first book *Divers voyages* of 1582 even his name does not appear; there is no portrait of Hakluyt.[19] He was buried in Westminster Abbey, but no inscribed stone covers his grave; it is not even known in which part of the church he lies. This modesty and self-renunciation are the marks of a true scholar. Sir Walter Raleigh, when Professor of English Literature at Oxford, once said:

> Over against the plays of Shakespeare and his fellows, as their natural counterpart, must be set the *Voyages* of Hakluyt; he who would understand the Elizabethan age, and what it meant for England must know them both . . . The dramatists and poets were the children and inheritors of the Voyagers.[20]

Hakluyt can be placed with his contemporaries Drake and Shakespeare as three of the greatest of the Elizabethans. Hakluyt died seven months after Shakespeare, on the 23 of November 1616. In the words of J. A. Williamson:

Just as Drake was not the only great seaman of his time, nor Shakespeare the only dramatist, so Hakluyt was not the only creator of the public pride in maritime achievement nor supplier of the facts upon which it rested. Each of these three in his own sphere was the greatest example of a widespread talent, its exponent, its moulder and its crown.[21]

Hakluyt's main object was to inspire Englishmen by recording their great maritime past. In Dr E. Lynam's words:

Hakluyt strove to show his countrymen that their future and the solution of their problems lay in maritime and colonial expansion. He devoted his life, not without success, to changing the outlook of a nation.[22]

Hakluyt's prose was worthy of its time. He wrote when English prose was at the stage of its most robust development. It behoves all English-speaking geographers first to re-read the work of Hakluyt; and then to reflect on the reasons why so much modern academic geographical writing in English or 'American' possesses neither clarity nor beauty of expression.

NOTES

1 This chapter was given as the first of four valedictory lectures on 'some Oxford geographers' at Oxford on 26 April 1967: it has not been printed before.

2 O'Meara, John J., *The First Version of the Topography of Ireland by Giraldus Cambrensis* (Dundalk, 1951), 6.

3 Ibid, 9.

4 Knowles, David, *Saints and Scholars. Twenty-five medieval portraits,* (Cambridge, 1962), 77. *The Itinerary Through Wales* and *The Description of Wales* were edited in Everyman's Library in 1908 by W. Llewelyn Williams.

5 Gunther, R. T., *Early Science in Oxford,* vol XI (1937), 42–3; Symons, G. J., *Consideraciones Temperiei pro 7 Annis, 1337–1344* (Reproduced and translated London, 1891).

6 Baker, J. N. L., 'Academic geography in the seventeenth and eighteenth centuries', *Scott Geog Mag,* 51 (1935), 5 for a brief account of Baldwin, Norton.

7 For the life of Hakluyt see Parks, G. B., *Richard Hakluyt and the English Voyages* (1930); Taylor, E. G. R., *Late Tudor and Early Stuart Geography 1583–1650* (1934), 1–39; Taylor, E. G. R., 'Richard Hakluyt', *Geog J,* 109 (1947), 165–74. Parker, John,

Books to build an Empire. A Bibliographical History of English Overseas Interests to 1620 (1965) is a valuable bibliographical aid to students of Hakluyt and his age.

8   *Scott Geog Mag*, 47 (1931), 321–2.
9   Girtin, Tom, 'Mr. Hakluyt, Scholar at Oxford', *Geog J*, 109 (1953), 208–12.
10 Hakluyt, Richard, *Principall Navigations* (1589), 'Epistle Dedicatorie'.
11 Girtin, Tom, op cit, 211.
12 Hakluyt, Richard, *Principall Navigations* (1589), 'Epistle Dedicatorie'.
13 A Poor Law Act was passed in 1572 and this was followed by numerous Proclamations against vagabonds.
14 Taylor, E. G. R. (ed), *The Original Writings and Correspondence of the Two Richard Hakluyts*, Hakluyt Society, 2nd Series, vol 77 (1935), 234.
15 The *Discourse of Western Planting* was reprinted by the Hakluyt Society in 1935. See Taylor, E. G. R. (ed), op cit, Hakluyt Society, vol 77, 211–326.
16 Lynam, E. (ed), *Richard Hakluyt and his Successors*, Hakluyt Society, vol 93 (1946), 35.
17 Taylor, E. G. R. (ed), op cit, Hakluyt Society, vol 77, 474.
18 Taylor, E. G. R., *Geog J*, 109 (1947), 165.
19 See Quinn, David B., *Richard Hakluyt, Editor* (1967).
20 Raleigh, Walter, *The English Voyages of the Sixteenth Century* (1910), 151.
21 Williamson, J. A., in Lynam, E. (ed), *Richard Hakluyt and his Successors* (1946), 14.
22 Lynam, E., op cit, 175.

CHAPTER TWO

'Geographie is Better than Divinity'

A T THE PRESENT day the academic study of geography grows ever more specialised and professional.[1] As a result it becomes increasingly difficult to combine the pursuit of geography with work in other fields of knowledge. In earlier centuries this was happily not the case and during the first sixty years of the seventeenth century several Oxford divines proved that 'geography was not incompatible with divinity'.[2] Preacher Richard Hakluyt had died in 1616, but he had many successors and the line of Oxford geographers continued unbroken during the next two centuries. In J. N. L. Baker's opinion the progress of academic geography in Britain during 'the seventeenth and eighteenth centuries is largely concerned with the University of Oxford which during that period led the way in geographical study and accomplishment, and produced one work of outstanding merit'.[3]

George Abbot was born at Guildford on 29 October 1562. He entered Balliol in 1579 and five years later became a probationer-fellow of the college. In 1597 he was elected Master of University College, Oxford, and during his tenure of that office wrote for his pupils a geographical treatise, *A Briefe Description of the Whole Worlde*, first published in 1599. Abbot was three times Vice-Chancellor of the University of Oxford, became successively Dean of Winchester, Bishop of London and, in 1611, Archbishop of Canterbury. In 1621, when with a shooting-party at Bramshill Park, he aimed at one of the deer, but his bolt struck Peter

44

Hawkins, Lord Zouch's gamekeeper, who bled to death in an hour's time. This unfortunate accident had unhappy consequences for the Archbishop, but in due course he received the King's pardon. Abbot has been described as 'stiffly principled in puritan doctrines', and he was in frequent conflict with Laud.[4] By geographers he is best remembered for giving patronage and encouragement to Samuel Purchas, that painstaking successor to Hakluyt. Abbot first made Purchas one of his chaplains and later gave him preferment to a living in Essex; Purchas dedicated *His Pilgrimage* (1613) to the Archbishop.

Abbot's *Briefe Description of the Whole Worlde* must have supplied a need, as it was constantly reprinted. A third and considerably enlarged edition appeared in 1608 and a sixth edition came out in 1624. The last (ninth or tenth) edition was printed in 1664, over thirty years after the author's death in 1633. Professor E. G. R. Taylor has described Abbot's book as 'this arid little compilation . . . for the most part a mere catalogue of place-names, forerunner of the dreary geographies which held the field in English schools right down to the twentieth century.'[5] This is not entirely just, as the book served a useful purpose in its day. J. N. L. Baker has pointed out that Abbot's treatise 'was doubtless intended to supply the latest geographical information for the benefit of those who were reading the standard works of ancient authors, and its general plan was copied by a number of writers'.[6] Abbot included an account of *America sive Orbe novo*. His remarks on *Russia sive Moscouia* are still apposite. 'This Empire', he says, 'is at this day one of the greatest dominions in the world: both for compasse of grounde, and for multitude of men; saving that it lyeth far North, and so yeeldeth not pleasure or good trafique, with many other of the best nations'.[7]

Peter Heylyn (Plate 1) was born at Burford on 29 November 1599 and attended the local Grammar School. In January 1615, at the early age of fifteen, he was admitted a commoner of Hart Hall, Oxford, to his 'great contentment'[8] six months later he was elected a Demy of Magdalen College, to which he migrated. 'After he had taken the degree of Bachelor of Arts which was in October 1617, he read every Long Vacation till he was Master, Cosmography lectures in the common refectory of Magdalen College, of which the first being performed in the latter end of July 1618, it was so well approved, that for that, and his other learning, he was chosen probationer, and in the year following perpetual fellow'.[9]

In 1621 he published his *Microcosmus: Or a little description of the great world*, which was presumably based on his lectures. He presented a copy personally to Charles, Prince of Wales, to whom it was dedicated, and by whom he 'was received very graciously'. In 1625 he travelled in France for about six weeks and in 1628 he went to Guernsey for three weeks as chaplain to the Earl of Danby. These two short journeys were recorded in a book which was not published until 1656. Heylyn was primarily a student of divinity; he took his BD in 1629 and DD in 1633. He wrote a large number of theological works, some of which involved him in bitter disputes; he was described as an 'acrimonious controversialist'.[10] Unlike Abbot, Heylyn was a high churchman and a strong opponent of puritan opinions. He was a devoted follower of the Royalist cause in the time of the Civil Wars, and lost everything including his library. In 1647 he was able to settle at Minster Lovell and returned from theology to the study of geography and history, which he had loved in his youth. During the following five years, in spite of failing eyesight, he composed his *Cosmographie*, a folio volume in four books, which was first published in 1652 and can be regarded as an enlargement of his *Microcosmus*. At the Restoration Heylyn became sub-dean of Westminster; he died on 8 May 1662 and was buried in Westminster Abbey.

Heylyn's first geographical work *Microcosmus* appeared in 1621 and was often reprinted; the eighth and last edition is dated 1639. The author described it as 'a treatise, historical, geographical, political and theological'. Certainly the book contains much miscellaneous information and is similar in character to Abbot's *Briefe Description*. Heylyn cites Ptolemy's definition of geography as 'an imitation of the picture of the whole Earth', before giving his own definition of the subject as 'a description of the earth, by her parts and their limits, situations, inhabitants, cities, rivers, fertilitie and observable matters, with all other things annexed thereunto'.[11] Heylyn's account of the two short journeys he made in his youth was published in 1656.[12] His description of France is confined to four provinces, Normandy, Ile de France, Beauce and Picardy. He says in his preface that, though he described these provinces only 'in the way of Chorography, yet I have took a general and *a full survey of the state of France,* in reference to the Court, the Church and the Civil State, which are the three main limbs of all Bodies Politick'. He justified his restricted account of France by describing the four provinces through which he had travelled as 'the epitome of the whole . . . the Isle of France being looked on as the mother of Paris, Picardie as the chiefest granary, and La Beauce as the nurse thereof; as Normandy is esteemed for the Bulwark of all France itself by reason of that large sea-coast, and well-fortified havens, wherewith it doth confront the English'.[13] Everywhere he was impressed with the dire poverty of the French peasants. 'Search into their houses', he writes, 'and you shall finde them very wretched, destitute as well as of furniture as provision. No Butter salted up against Winter, no powdring tub, no Pullein in the Rick-barten, no flesh in the pot or at the spit, and which is worst, no money to buy them'.[14] His account of the Channel Islands is of no great interest; he regarded them as 'seated purposely for the command and Empire of the ocean'.[15] Heylyn was criticised by Phillippe Falle, whose excellent *Account of the Island of Jersey* appeared in 1694; 'Dr. Heylin is he who wrote the fullest [account of Jersey]. But it was not possible that in one Week he should gain any other than a very superficial knowledge of our Affairs'.

John Evelyn heard Peter Heylyn preach in Westminster Abbey on Friendship and Charity on 29 March 1661 and in his diary he records that the preacher, 'the author of the Geography', was quite blind and had been so for some years. Heylyn's fame as a geographer clearly rests on his *Cosmographie.* which he wrote at Minster Lovell with his own hand and was first published in 1652. After that date his eyes failed him and he could neither read nor write without the help of an amanuensis. The book was frequently reprinted after its author's death; the third edition appeared in 1666, the sixth in 1682 and in 1703 a new edition 'improv'd with an historical continuation to the present time' was made by E.

47

Bohun. Heylyn's *Cosmographie* was a popular standard work in its time, but it is not an easy book to read today. Professor E. G. R. Taylor devotes several pages to Heylyn and ascribes his 'failure to satisfy a geographical reader' to the fact that 'he neither felt disposed, nor was he competent to give to geographical description the space it deserved'.[16] She add that 'he awakens interest, but fails to satisfy it'.

Heylyn's book is a mixture of history and geography and the author makes some sound remarks on the relation between the two subjects. '*Geography* without *History*,' he writes, 'hath life and motion, but very unstable and at random; but *History* without *Geography*, like a dead carkasse, hath neither life, nor motion at all.' He concludes, with some confusion of image, that history and geography, 'if joyned together, crown our reading with delight and profit; if parted threaten both with a certain shipwrack: and are like two Sisters dearly loving, not without pity (I had almost said impiety) to be kept asunder'.[17] Heylyn's volume was based largely on secondary authorities and not on his own observations or even on those made by travellers. J. N. L. Baker believes that Heylyn made some use of the work of Nathanael Carpenter, Fellow of Exeter College, whose valuable *Geographie Delineated Forth in Two Bookes* was published at Oxford in 1625.[18] Heylyn certainly made much use of Keckermann's writings as well as those of Jean Bodin, S. Du Bartas and other Continental authors.

In an appendix to his book Heylyn gives some account of 'the unknown parts of the world, especially Terra Australis Incognita or the Southern Continent'. He adds that 'he will make a search into this *Terra Australis* for some other Regions, which must be found either here or no where'.[19] The seven regions in this category included Utopia, New Atlantis, Fairy Land, the Lands of Chivalry and the New World in the Moon. Of the last named he wrote: 'that world . . . is become a matter of *serious debate*: and some have laboured, with great pains, to make it probable that there is another *World* in the *Moon*, inhabited as this is, by persons of divers *Languages*, *Customs*, *Polities*, and *Religions*: and more than so, some means and wayes proposed to consideration for maintaining an *intercourse* and *commerce* betwixt that and this'.[20]

Occasionally Heylyn gives accounts of incidents that concerned himself. These are illuminating and prove how much his geographical work would have benefited by more personal experience of travel. One such passage vividly reveals the troubles brought by political and religious controversy to the men of seventeenth-century England:

> And here I cannot but remember a pretty accident which befell me in the moneth of *January*, An. 1640. at what time it had been my ill fortune to suffer under some misapprehensions which had been entertain'd against me, and to be brought before the *Committee for the Courts of Justice*, on the complaint of Mr. *Prynne*, then newly return'd from his confinement, and in great credit with the Vulgar. Heard by them, I confess, I was, with a great deal of ingenuous patience; but most despightfully reviled and persecuted with excessive both noise and violence, by such as thronged about the doors of that *Committee*, to expect the issue: it being as naturall to many weak and inconsiderate men, as it is to Dogs, to bark at those they do not know, and to accompany each other in those kinds of clamors. And though I had the happiness to come off clear, without any censure, and to recover by degrees amongst knowing men, that estimation which before had been much endangered, yet such as took up matters upon trust and hear-say, looked on me as a person forfeited and marked out for ruin. Amongst others, I was then encountred in my passage from *Westminster to Whitehall*, by a tall big Gentleman, who thrusting me rudely from the Wall, and looking over his shoulder on me in a scorning manner, said in a hoarse voice these words, *Geographie is better than Divinity*; and so passed along. Whether his meaning were, that I was a better *Geographer* than *Divine*; or that *Geographie* had been a Study of more credit and advantage to me in the eyes of men, than *Divinity* was like to prove, I am not able to determine. But sure I am, I have since thought very often of it, and that the thought thereof had its influence on me, in drawing me to look back on those younger studies, in which I was resolved to have dealt no more: and

thereto, in the Preface to my *Microcosm*, had obliged my self.[21]

It is certain that after his unhappy misfortunes in the Civil War Heylyn turned once again to his old loves, the study of geography and history; and the five years of work which he spent in composing his *Cosmographie* must have given him great solace.

It has been noted that Heylyn was first a member of Hart Hall. Geographical studies in the seventeenth century were pursued at both Hart Hall and Magdalen Hall, of which two foundations the present Hertford College is the heir and successor. Thus the modern link between the University of Oxford's School of Geography and Hertford College merely continues an old tradition.[22] Both Hart Hall, founded in the thirteenth century, and Magdalen Hall, established at the end of the fifteenth, were nurseries of geographers. In view of the geographical labours of these men it is not surprising that the old library of Magdalen Hall, now a part of Hertford College library, includes much geographical literature. The formative years of the collection lay between 1650 and 1775.[23] Both Halls produced geographers of distinction in that period; two can be regarded as important pioneers in the history of geography. The two men were contemporaries and their geographical achievements were made at the same time. Robert Plot of Magdalen Hall was an Oxford pioneer in the growth of regional geography: John Caswell of Hart Hall was the first in Britain to measure mountains with precision, thus making an immensely valuable contribution to cartography.

Although he can hardly be claimed as a geographer, the philosopher Thomas Hobbes, one of the most distinguished graduates of Magdalen Hall, was certainly interested in maps and travel. He entered the Hall as a boy of thirteen in about 1602 and took his degree in 1608. He seems to have found the formal courses in logic and scholastic philosophy very repellent. So he turned to other studies which were more congenial to him, such as astronomy and the reading of books of travel. Aubrey says of him that while at Oxford 'he tooke great delight there to goe to the bookebinders' shops, and lye gaping on mappes';[24] and according to the same author, Hobbes also amused himself with snaring jackdaws.

Plate 3 (a) The south side of Mahón harbour. The former British quay for warships and the old naval storehouses can be seen

Plate 3 (b) Monument to Governor Richard Kane in Menorca. It is placed alongside the road, opened by Kane in 1720, which crosses the whole island from Mahón to Ciudadela

Plate 4 (a)   A house in Mahón with the English-type sash window

Plate 4 (b)   Arcaded street in Ciudadela

Hobbes wrote an account of the Peak district in Derbyshire and describes its limestone caverns, swallow holes, and lead-mines.[25] But his geographical interests were human as well as physical. In his *Leviathan* Hobbes expressed opinions about towns that are strongly held by some geographers of today. His view that 'one infirmity of a Common-wealth is the immoderate greatness of a Town'[26] is the theme of much writing by modern geographers and planners. What would Hobbes have said of the six conurbations of today now holding over 40 per cent of the population of England?

William Pemble (1592–1623) was another pious and learned Puritan divine who, like Abbot, contributed to geographical learning. He was admitted to Magdalen College in 1610, and, after taking his bachelor's degree in 1613, became 'a noted reader and tutor' at Magdalen Hall, and was described as 'an ornament to the society among whom he lived'.[27] He was a learned exponent of Calvinism and highly regarded as a preacher. Pemble died at the early age of thirty-one; all his written works, mostly theological, were published after his death. In 1630 his only geographical work was published at Oxford, *A Brief Introduction to Geography*, but it reached a fifth edition in 1675 and was again issued ten years later. The writer of the preface hoped that the book might be an introduction for younger students, 'for others a Remembrancer, for any not unworthy the perusall'. In his first chapter Pemble distinguished between topography, chorography and geography. 'Topographie', he wrote, 'is a particular description of some small quantity of Land, such as Land measurers sett out in their plotes. Chorography is a particular description fo some Country, as of England, France, or any shire or province in them: as in the usuall and ordinary mappe. Geography is an art or science teaching us the generall descrption of the whole earth.' He added that both chorography and geography were 'excellent parts of knowledge in them selves, and affording much profit and helpe in the understanding of history and other things.' Pemble next made a broad division of geography into 'generall, which treateth of the nature, qualities, measure with other general properties of the earth'; and 'speciall, wherein the several countries of the

D   53

earth, are divided and described'. He firmly asserted that 'the earth resteth immoveable in the very midst of the whole world'. This conservative geographer still held to the Ptolemaic theory nearly a hundred years after the death of Copernicus. He states that the 'opinion of Copernicus and others . . . is most improbable and unreasonable; and rejected by the most'.

Robert Plot (1640–96), the first Keeper of the Ashmolean Museum, matriculated from Magdalen Hall in 1658 (Plate 2). Although he must be regarded as a pioneer in the development of regional geography it is not necessary to outline his work in detail as this has been admirably done by F. V. Emery.[28] In 1674 he drew up 'a list of queries to be propounded to the most ingenious of each County' in his travels through England. Three years later he published his *Natural History of Oxfordshire* (1677), and this was followed by a *Natural History of Staffordshire* (1686), in which he divided the county into three natural units. Both books incorporated many observations made in the field; the latter is perhaps more attractively written. Plot asked questions about the air, waters, springs, navigation, earths and minerals, stones and quarries, metals, plants, husbandry, animals, arts and antiquities. Among his questions were the following: 'what diseases are most common here, or are there any peculiar to the place?'; 'how lye the beds of mould, clay, sand, etc. one above another?'; 'how many sorts of plows, carts, harrows, rolls, rakes, forks?'; and 'how is the name of this town or parish written?' Even in the seventeenth century the field worker, with his list of questions, had to reckon with the frailty of human nature, and Plot was very credulous. It is said that for many years afterwards it was a boast among the Staffordshire squires to whom he addressed his inquiries, how readily they had 'humbugged old Plot'.[29] The author of the two natural histories had a genial and discursive mind. He was a witty, bon-vivant Tory. He became Secretary of the Royal Society and, in 1682, founder of the Oxford Philosophical Society, to which four years later John Caswell of Hart Hall communicated his experiments in measuring the heights of mountains. So the story in geographical pioneering must move from Plot to Caswell.

Among the several distinguished geographers of Hart Hall was Richard Holland (1596–1677) who taught the use of the globes and other instruments (including Gunter's Quadrant), as well as mathematics,[30] He was succeeded, as tutor of mathematics, by John Caswell (1656–1712), who became Vice-Principal of the Hall.[31] Caswell was associated with John Adams in an attempt to make a geodetic survey of England and Wales in 1681–4.[32] The two men measured a base-line twelve miles in length in Somerset. Caswell must be regarded as the pioneer in the precise measurement of the heights of mountains in England and Wales, by using a mercury barometer, as had been attempted in France on the Puy de Dôme. Caswell was the first man to determine the height of Snowdon; on 17 July 1682, by means of a mercury barometer ('Torricellian tube', invented in 1643), he calculated that the Welsh mountain's summit was 1,240yd above the level of the sea. Caswell's result is only 159ft higher than the figure given on modern Ordnance Survey maps (3,561ft). He also measured other heights in Wales and Shropshire; his figure of 2,910ft for Cader Idris, observed on 26 July 1682, is only seventeen feet below the OS figure of 2,927ft.[33]

Edward Leigh (1602–71), a Magdalen Hall man, who took his degree in 1620, was a 'laborious student of divinity and law' as well as of geography. His theological works are somewhat arid but, like Robert Plot, he believed that geographical writing should be based on travel and not only on work in the study. As a result his *Guide for Travellers in Forein Parts* (1671) has especial interest.[34] He quotes with approval the remark that 'there is no map like the view of the country; one journey will shew a man more than any description can'. Leigh's book was a forerunner of the modern *Hints to Travellers*, published by the Royal Geographical Society, with its catalogues of items which must be observed and recorded. Leigh gives full details of foreign money and of the many different measures of distance between place and place. He advised that all travellers should 'have the Latin tongue', and 'be skilful in architecture, able so well to limn or paint, as to take in paper the situation of a castle, or a city.' Leigh also advised every traveller 'to inform himself by the best Chorographical and

Geographical Map of the Situation of the country he goes to, both in itself and relatively to the Universe, . . . and to carry with him a Map of every Country he intends to travel through'. His final instruction to the traveller was that before his voyage 'he should make his peace with God, receive the Lord's Supper, satisfy his creditors if he be in debt,' and 'make his last will, and wisely order all his affairs, since many that go far abroad, return not home'.

The Turners, father and son, both Magdalen Hall men, carried on the hall's geographical tradition during the eighteenth century. Richard Turner the elder (1724–91) matriculated in 1748 and became another parson who combined divinity and geography. He was ordained and held several livings including Comberton and Elmley, and taught mathematics and philosophy at Worcester. He was the author of *A View of the Earth: being a short but comprehensive system of modern geography* (1762). His second son, Richard Turner the younger (1753–88), matriculated from Magdalen Hall in 1773. He wrote several geographical works of which *A new and easy Introduction to Universal Geography* (1780) achieved great popularity, reaching its thirteenth edition in 1808. It was of an elementary character and written in the form of a series of letters. He, too, had combined geography with divinity, for his earliest work, *An Heretical History* (1788) was a compilation in which he set forth the doctrines of the various heretical sects of the early Christian world.

NOTES

1 The greater part of this chapter was first published as 'Geographie is better than divinity' in *Geogr J*, 128 (1962), 494–7; extracts have also been taken from the author's 'Geography at Hertford College', *Hertford Coll Magazine* (May 1958), and his Inaugural Lecture, to the University of Oxford, *Geography as a Humane Study* (1955).

2 Taylor, E. G. R., *Late Tudor and Early Stuart Geography 1583–1650* (1934), vi.

3   Baker, J. N. L., 'Academic geography in the seventeenth and eighteenth centuries', *Scott Geogr Mag*, 51 (1935), 129; and 'Nathanael Carpenter and English geography in the seventeenth century', *Geogr J*, 71 (1928), 261–71.

4   *Dictionary of National Biography*. See also Welsby, Paul A., *George Abbot. The Unwanted Archbishop 1562–1633* (1962).

5   Taylor, E. G. R., op cit (1934), 37. In the opinion of Dr Paul Welsby, Professor Taylor's judgement of the book was unfair, because 'Abbot was pioneering in a new field'.

6   Baker, J. N. L., op cit (1935), 131. Abbot was deeply interested in places overseas; in later life he was associated with the Adventurers to Virginia in which he held 75 shares (Welsby, op cit).

7   Abbot, George, *A Briefe Description of the Whole World* (2nd ed, 1600), 5. The first edition is dated 1599. The book has a curious appendix, a list of the universities of the world with the latitude and longitude of each.

8   Bloxham, J. R., *A Register of the Presidents, Fellows, Demies of Saint Mary Magdalen College, Oxford* (1876), vol 2, 50. This contains on pp 48–57 a full account of Heylyn's life with extracts from his diary. See also Vernon George, *The Life of Dr. Peter Heylyn* (1682).

9   Bloxham, J. R., op cit, vol 2, 58; Wood, A., *Athenae Oxonienses* (ed Philip Bliss, 1817), 552.

10 *Dictionary of National Biography*.

11 Heylyn, Peter, *Microcosmus; Or a little description of the great World* (1621), 10.

12 Heylyn, Peter, *A Survey of the Estate of France and of some of the Adjoining Islands* (1656), in five books. The sixth book is *The Second Journey: A Survey of the Estate of the two Islands, Guernsey and Jersey, with the Isles appending.*

13 ibid, preface.

14 ibid, 259.

15 ibid, 293.

16 Taylor, E. G. R., *Late Tudor and Stuart Geography*, op cit, 142.

17 Heylyn, Peter. *Cosmographie* (1657), 19. Quotations are taken from the second edition of 1657. The first edition is dated 1652.

18 Baker, J. N. L., op cit, *Geogr J*, 71 (1928), 269–70

19 Heylyn, Peter, *Cosmographie*, op cit, appendix, 1093.

20 ibid, appendix, 1095.

21 ibid, preface, 2.

22 In 1932 a chair of Geography was instituted at Oxford and was attached to Hertford College under the scheme of allotment of professorship to colleges. The first three holders of the chair

automatically became Fellows of Hertford on their election to the professorship.

23 For a brief account of the geographical section of the old library of Hertford College, see *Victoria County History, Oxfordshire,* vol III (1954), 315.

24 Aubrey, John, *Brief Lives* (ed, Clark, Andrew, 2 vols 1898), I, 329.

25 Hobbes, Thomas, *De Mirabilibus Pecci: being the Wonders of the Peak of Derbyshire* (1636). See Taylor, E. G. R., *Late Tudor and Early Stuart Geography 1583-1650* (1934), 147 and 154-5.

26 Hobbes, Thoms, *Leviathan* (1651), chapter 29.

27 Wood, A., *Athenae Oxonienses* (ed, Bliss, Philip, 1813-20), vol 2, 335.

28 Emery, F. V., 'English regional studies from Aubrey to Defoe', *Geogr J,* 124 (1958), 315-25.

29 *Dictionary of National Biography.*

30 Richard Holland was the author of *An Explanation of Mr. Gunter's Quadrant* (1676) and *Globe Notes* (1678).

31 In New College Lane, Oxford, only a few yards from Hart Hall, John Prujean (died 1701) made and sold mathematical instruments, including Quadrants designed by Holland and Nocturnals by Caswell. See Taylor, E. G. R., *The Mathematical Practitioners of Tudor and Stuart England* (1954).

32 Taylor, E. G. R., 'Robert Hooke and the cartographical projects of the late seventeenth century 1666-1696', *Geogr J,* 90 (1937), 529-40.

33 Other hills measured by Caswell between July and September 1682 were the Wrekin, 1,398ft (OS 1,335ft); Penmaenmawr, 1,545ft (OS 1,550ft); Clee Hill, 1,800ft (OS 1,750ft); and Stiper Stones. A letter from Caswell with an account of his observations of the heights of mountains and hills was read at a meeting of the Royal Society on 30 June 1686.

34 Edward Leigh wrote two works of a geographical nature, the first being *England Described, or the several Counties and Shires thereof briefly handled* (1659), mostly taken from Camden's *Britannia* (1587). The second appeared in 1671, the year of the author's death, under the title *Three Diatribes or Discourses. First of Travel, or a Guide for Travellers into Forein Parts, Secondly, of Money or Coyns. Thirdly, of Measuring of the Distance between Place and Place.*

CHAPTER THREE

Influences of British Administration on the Human Geography of Menorca

MENORCA (MINORCA) IS one of the Balearic group of islands, and as it lies over twenty miles north-east of Mallorca, it is the most easterly part of Spain.[1] The island is about thirty miles long from east to west and averages between eight and twelve miles in breadth; its area of 258 square miles is approximately the same as the area of the Isle of Man, and is only about one-fifth the size of Mallorca. It forms the north-eastern extremity of the Balearic submarine ridge, and its northern coasts overlook deep water. The average height of Menorca is considerably less than that of Mallorca, as there are no mountains comparable to the north-western Sierra of the greater island. Menorca is, in fact, very different in character from Mallorca in many other ways than in mere size. The principal axis of the island is at right angles to the main direction of the group of islands as a whole, and is roughly along a line from Mahón to Ciudadela. This line divides the island into two unequal portions. The smaller northern region, is largely composed of rocks of Palaeozoic age, principally Devonian, although there are some areas of Jurassic limestone: the Devonian rocks are not found elsewhere in the Balearic archipelago.[2] The southern region, which covers about two-thirds of the total area of the island, is composed of horizontal layers of limestone of middle Miocene. The northern region is more rugged in relief than the southern, but its hills are low and discontinuous, the highest, Monte Toro, in the centre of the island, being only 1,175 ft above sea-level. This hill was used as a watch-post by the British troops during the occupation, as the whole island can be seen from its summit. The southern region is a low, undulating plateau, whose general level is between 150 and 350ft. This tableland is broken by deep gorges, which run in a general north to south direction, and have fertile bottoms; the south coast has steep cliffs. The southern region is somewhat similar in character to the central plain of Mallorca, but the northern region may be described as more Catalan than Balearic in its relief and structure. It has been said that Menorca should be regarded as 'a piece of débris from the continental massif which formerly united Sardinia with Catalonia'. This division of Menorca into two regions has been of great significance from the earliest times. A map of the distribution of the numerous prehistoric remains shows that almost all of them are situated in the southern region, which is still today the region of greatst human activity (Fig 1).

The low relief of Menorca, combined with its small size and more northerly position, has given it a climate distinct from that of the other islands of the Balearic group.[3] The whole of Menorca is very open to the influence of the sea, and it is certainly wetter and milder than Mallorca. The low relief leaves Menorca exposed to the ravages of all the winds of heaven, and it might well be called the island of winds, in contrast with its neighbour Mallorca, the island of calm. The range of temperature is 2° F less than that of Palma. (Mahón 51° Jan, 76° Aug; Palma 51° Jan, 78° Aug.) The mean annual rainfall of Mahón is over 26in, which is 7in greater than the corresponding figure for Palma. Mahón has, on the average, 78·8 rainy days against Palma's 73. Of these 78·8 rainy days, 27·3 are in September-November, 24·6 in December-February, 18 in March-May, and 7·9 in June-August. Two winds predominate in Menorca, namely, the south-west (*Llebeitx* or *Llebeig*) and the north (*Tramontana*). The former is accompanied by rain, the latter by cold. If the *Tramontana* is made to include north-west and north-east winds, it may be said to blow on 165 days of the year. It is generally a strong, dry wind, and on 105 of

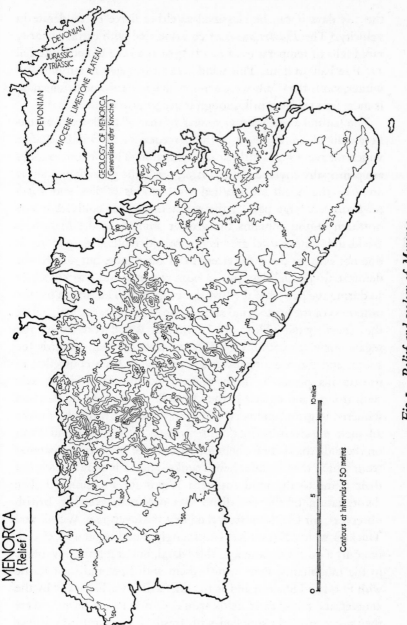

MENORCA
(Relief)

DEVONIAN

DEVONIAN

JURASSIC
TRIASSIC

MIOCENE LIMESTONE PLATEAU

GEOLOGY OF MENORCA
(Generalised after Knoche)

Contours at intervals of 50 metres

0   5   10 miles

Fig 1 Relief and geology of Menorca

the 165 days it can be classified as either fierce or moderate in velocity.[4] The *Tramontana* is often associated with extraordinarily rapid falls of temperature. On 17 April 1926 the temperature fell 11° F in half an hour. This wind is most frequent in autumn and winter, while the south-west wind, which is warm and brings rain, is most frequent in April, though it occurs also in the winter.

The natural vegetation is similar to that of Mallorca, but it is very stunted, as the result of the fierceness of the winds, against which the relief gives little or no protection. The Devonian rocks are principally covered with *garrigue*, while the Jurassic limestone areas of the north are clothed with woods of ilex and *Pinus halepensis*.[5] A large part of the southern plain is cultivated. The northern region contains the greater amount of the Menorcan woodland, whose total area is not considerable. This poverty is due not only to the great strength of the winds, but also to the deforestation which has always been reckless, and was especially so during the period of the British occupation. The British felt the bitterness of the north wind and, as one observer remarked, 'while they (north winds) blow, we lay on our Wood very freely, and regale ourselves within Doors'.[6] The trees are bent towards the south, and the Menorcan farmer has always been compelled to protect his plants by high walls. 'The Mountain-Pines', said Armstrong, 'and especially the Olive-Trees, are ever stunted and withered in great Numbers, and even those that thrive best make all their vigorous Shoots, and produce their Cones and Fruit on the Side that is best sheltered from those blustering Enemies, from which they incline their Trunks, and as it were stretch out their Arms to the mild southern Winds for Protection'.[7] The deforestation of the woodland was deplored, even by British observers, for Cleghorn remarked that the 'evergreen Woods and Thickets which Nature has surprizingly raised upon a Rock, are not only a great Ornament to this Island, but of infinite advantage to the Inhabitants; they furnish them with Fuel and their Cattle with Food and Shelter; and as the finer soil is washed away by the anniversary rains, their fields would soon become barren, were they not constantly supplied with fresh Manure from the leaves of the Vegetables.... The natives therefore are much to be blamed in selling so many of their trees, and stubbing up the roots so rashly, as they have done of late Years, for immediate Profit, since the damage will soon be sensibly felt and not easily retrieved by their Posterity'.[8]

Menorca is a bleak and sterile land, and fully justifies the name given to it by a Spanish novelist as 'the island of stones and winds.' It is surprising that such a barren place can support a population of over 40,000. It must be noted, however, that the density of population per square mile was 161 in 1930, which was lower than the density of Mallorca (200). Agriculture is the principal occupation, but the life of the Menorcan farmer is generally more difficult that that of the Mallorcan. The violence of the winds makes the cultivation of fruit trees, the main source of Mallorca's wealth, almost impossible in Menorca. Rather less than half the island was cultivated in 1934 (46 per cent, ie 30,050 hectares). Woodland covered 33 per cent (22,400 hectares), while 19 per cent (13,000 hectares) was entirely unproductive.[9] The numerous walls which divide the southern plain into enclosures called *tancas* were one of the most noticeable features of the scenery. These walls are due to the abundance of stone which must be cleared from the fields in order that cultivation of any kind may take place. The enclosures enable flocks and herds to be pastured without men to watch over them. They also facilitate the alternation of cultivation and pasture, and to some extent protect crops from the wind.

There are no rivers in Menorca, but the numerous springs are used for irrigation. The total irrigated area is very small (1 per cent of the whole in 1934, ie 550 hectares), the most important region being the *huerta* of San Juan, originally drained by the British Governor, Kane.

Menorca has not been richly endowed with natural gifts, and it seems strange, at first sight, that so much importance was attached to a poor and thinly populated island during the eighteenth century. The reasons for the prominent part played by the island during that century was purely stratetgic. The population has been roughly estimated at about six thousand in the year 1588, and as between fourteen and fifteen thousand in 1700. One of the objects of the present chapter is to show that some of the peculiarities of the distribution of population in Menorca at the present day are the results of the military occupation of the eighteenth century. During the greater part of that century Menorca did not belong to Spain, but to foreign rulers. This island lies in the centre of the western basin of the Mediterranean Sea, almost due south of the mouth of the Rhône, and it was this fact which gave it such great significance in naval strategy. This important was much greater before the opening of the Suez Canal. As Menorca is only 240 miles from Toulon, it would obviously form a useful base for a British fleet operating against the French, provided that it contained a good harbour. In actual fact, Menorca contains one of the best harbours in the Mediterranean for sailing vessels. A Spanish proverb runs:

> 'Los puertos del Mediterráneo son
> Junio, Julio, Agosto y Puerto Mahón.'

The entrance to the natural harbour of Port Mahón is only 250yd wide, but it extends inland for 6,000yd, and is between 500 and 1,000yd wide in places. The existence of this fine harbour made Menorca a very desirable possession for a naval power.

Menorca was taken by the British in 1708, and held until 1756. During the first British occupation three English books were written about the island. The first, by Colin Campbell, was published in 1716 and is entitled *The Ancient and Modern History of the Balearick Islands*. This book is largely a translation of Spanish works by Dameto and Mut, but it contains a brief appendix on 'The reducing of these islands by the arms of Great Britain'. The second book was by George Cleghorn, Lecturer of Anatomy in the University of Dublin, formerly Surgeon to the Twenty-second Regiment of foot, and is entitled: *Observations on the Epidemical Diseases in Minorca from the year 1744 to 1749, to which is prefixed a short account of the Climate, Production, Inhabitants and Endemical Distempers of that Island*. This work was published in 1751, and new editions appeared in 1761 and in 1768; it was dedicated to the Society of Surgeons of His Majesty's Royal Navy. The author kept a diary of the weather and correlated his meteorological records with the incidence of diseases, and he hoped that his observations would be of assistance to British surgeons whose work took them to regions with a climate similar to that of Menorca. Cleghorn describes his work as 'an Account of the Diseases only of a small, remote Part of the British Dominions; but of a Part in which Numbers of His Majesty's subjects, besides the Natives and those employed in the Protection of the Place, are often brought together, both in Time of Peace and War: And as the Qualities of the Air, and the Course of the Seasons in Minorca correspond nearly with those in several other parts of the World, to which our Fleets frequently repair, it is probable the Diseases may likewise be similar'. The third and most important book was written by John Armstrong, who was ordered to Menorca in 1738 as Engineer in Ordinary to His Majesty. It was first published in 1752 under the title of *The History of the Island of Minorca*. This book is written in the form of a series of letters, and also contains a map and several illustrations; it is the most valuable printed source for reconstructing the geography of Menorca in the eighteenth century. The work was later issued in German, French and Spanish translations. Armstrong explained that his object was 'to withdraw the veil that has so long hid these islanders from the observation of their neighbours, and continued them, though they make a part of our British Dominions, as utter strangers to the good people of England, as the Hunters of Aethiopia, or the Artificers of Japan'.[10]

In the name of the Archduke Charles, an English fleet under Sir John Leake landed a combined force of 2,000 British, Portuguese and Spanish troops south of Cala Alcaufa (on the east coast of Menorca) on 14 September 1708 (Fig 2). Great difficulty was experienced in landing the artillery of 42 cannon and 15 mortars at Alcaufa, but Fort St Philip, garrisoned by 1,000 men, was captured with not more than fifty British casualties on September 28. Two British ships, the *Dunkirk* and the *Centurion*, battered down Fort Fornells on the north coast, and a detachment of foot marched across the island to capture Ciudadela. The British occupation of Menorca was recognised by the Treaty of Utrecht in 1713. Armstrong says that it was 'a very valuable acquisition on every account, especially because of its excellent port, which

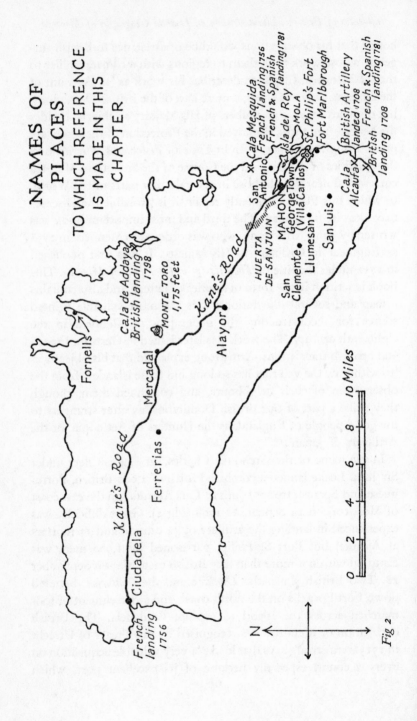

NAMES OF
PLACES

TO WHICH REFERENCE
IS MADE IN THIS
CHAPTER

Fornells

Cala de Addaya
British landing
1798

MONTE TORO
1,175 feet

Mercadal

Kane's Road

Kane's Road

Ciudadela

French landing
1756

Ferrerias

Alayor

Cala Mesquida
French landing 1756
French & Spanish
landing 1781

San
Antonio

Isla del Rey

Isla del Rey

MOLA

St. Philip's Fort

Fort Marlborough

British Artillery
landed 1708

HUERTA
DE SAN JUAN

MAHON

George Town
(Villa Carlos)

San
Clemente

Llumesan

San Luis

Cala
Alcaufar

French & Spanish
landing 1781

British landing 1708

N

0   2   4   6   8   10 Miles

Fig 2

immediately became the ordinary Rendezvous of the British fleet in the Mediterranean. For here they could assemble as many ships as they pleased in the utmost secrecy, without having their destination discovered to the enemy, as was likely enough to happen in the harbours of the Continent. Here too they were a kind of curb on the French and other maritime powers. But among the several advantages that redounded to the British nation from this Conquest, it raised their figure in these parts, and gave the Moors and the Italian States an idea of them more favourable than they had entertained before, and rendered them more tractable in certain negotiations that ensued, than they would possibly have been, if this enterprize had not been crowned with success'.[11] Campbell remarked that the port of Mahón 'is capable of being made of very great use in protecting the commerce of Great Britain in the Mediterranean, a trade of so great consequence to the Nation'.[12]

In 1712, Colonel Richard Kane (Plate 5) became Lieutenant-Governor of the island, and held that office or that of Governor until his death in 1736, except for a period of two years when he was absent as Governor in Gibraltar.[13] Kane is one of the earliest English examples of the enlightened colonial administrator, and his actions in Menorca are worthy of consideration. The island continued to enjoy its old laws and the free exercise of its religion 'to the no small satisfaction of their priests and lawyers'.[14] Kane began, like many colonial governors after him, by building roads. A road had been built by the Romans between Ciudadela and Mahón, but it was scarcely passable in 1700. Kane ordered the construction of a new road from Fort St Philip to Ciudadela in 1713, but it was not opened for public use until 1720. (Fig 2). The greater part of this road is over 30ft in width; some of it is still in use, but the general line was superseded by a new main road built in the years 1878–1900. The march from Mahón to Ciudadela took two days, and when the British regiments changed their stations, which they did every year in April or May, they halted for one night at Mercadal.[15] The construction of Kane's road necessitated the drainage of the marshes at the head of Mahón harbour; a considerable extent of ill-drained land was thus converted into the fruitful *huerta* of San Juan and divided into plots. This is still the largest area of irrigated land in the island, and a monument to Kane stands, very appropriately, on the edge of his road and overlooks the *huerta* (Plate 3 (*b*)). One of the most important acts of Governor Kane was to transfer the capital. The old capital and seat of the courts of justice was Ciudadela, situated at the western end of the island; it had been the capital since the Moors surrendered to King Jaimé in 1232. Kane, by a decree issued in 1721, ordered the courts of justice to be moved to Mahón, and took up his residence as Governor in that town. The change was made because Port Mahón was obviously a much better naval harbour than Ciudadela. The harbour of the old capital was small, but it was nearer Mallorca and the mainland than Mahón, and had served 'very well for the barks that traded to Majorca and the Continent, and it supplied Mahón with foreign goods'.[16] The removal of the seat of the capital to Mahón had a disastrous effect on the prosperity of Ciudadela. At the time of the British conquest the two places had about the same number of inhabitants. Both were still surrounded by their medieval walls, and very few houses had been built outside these defences. As soon as the

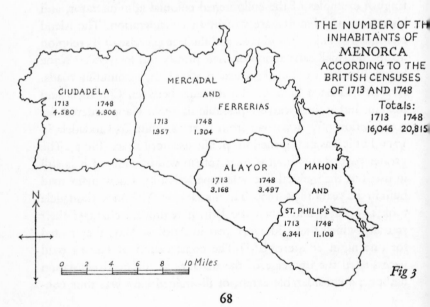

THE NUMBER OF TH[E]
INHABITANTS OF
MENORCA
ACCORDING TO THE
BRITISH CENSUSES
OF 1713 AND 1748

CIUDADELA
1713   1748
4.580   4.906

MERCADAL
AND
FERRERIAS
1713   1748
1.957   1.304

Totals:
1713   1748
16,046   20,815

ALAYOR
1713   1748
3.168   3.497

MAHON
AND
ST. PHILIP'S
1713   1748
6.341   11.108

N

0 2 4 6 8 10 Miles

Fig 3

change was made, 'Ciudadela visibly declined in its trade, and its wealth decreased in the same proportion; and the number of its inhabitants suffered, by swift degrees, a very sensible diminution'.[17]

Kane realised that statistical information was essential for the proper administration of the island. Henry Neal, HM Surveyor of Minorca, included a census of population, dated 20 April 1713, in his 'State of the Island of Minorca, 1713'.[18] The population of the island was then given as 16,046, excluding the garrison and the clergy. There were 4,301 men of military age (16–20) (Fig 3). In 1717 another enumeration provided additional statistical information by listing the number of farms, animals, wells and inhabited houses as well as rents.[19] Another census, taken by General William Blakeney in 1748 showed that the island's population had increased to 20,815,[20] (Fig 3). The increase of 4,769 at the end of thirty-five years of British rule was almost entirely concentrated on Mahón. Ciudadela had increased by only 326, Alayor by 329, while Mahón had increased by 4,767 (Mahón by 3,495, and St Philip's by 1,272). Mercadal (including Ferrerias and Fornells) had declined by 653. Both censuses omitted from their enumeration the clerical population (about 300 in number) and the garrison. John Armstrong estimated the total population of Menorca in 1756 as nearly 28,000. This may have been an over-estimate, but it must have included the garrison (about 2,400) and the clergy. The growth of Mahón's population was of course accompanied by extensive building operations. In 1756 Armstrong estimated that the new capital contained between 800 and 1,000 houses as against Ciudadela's 600. The ancient walls of Mahón were partly destroyed and the subsequent expansion of the town was carefully planned by Kane. New streets were built and old ones were widened. Some of Kane's names, such as Hanover Street and George Street, still survive in the modern town. A new road was built down to the harbour; the Governor's house was enlarged. The harbour itself was improved by the construction of a long quay below the town, the eastern half being reserved for naval vessels and naval stores, while the western half was given to merchant shipping. (Plate 3 (*a*).) A naval dockyard and a store for masts were built on the north side of the harbour on a site that had formerly been an unhealthy swamp, while the harbour itself was adequately charted. The Fort of St Philip was rebuilt and honeycombed with underground passages and galleries; the immense sum of one million pounds was spent on its reconstruction, so that it became known as 'the second Gibraltar', and has been described by a recent historian as 'one of the finest examples of military engineering in Europe',[21] The inhabitants of the island were 'vastly enriched by the immense sums that have been sent into their Country, for the payment of the troops, and for the works that have been erected'.[22]

Kane also concerned himself with the economic life of the islanders. In 1724 he standardised weights and measures and made the use of weights that were imported from London compulsory throughout the island. Menorca was, and is, a poor island agriculturally. As great difficulty was experienced in obtaining sufficient food for the garrison and the navy, Kane introduced various measures that were designed to improve and increase supplies. 'When he (Kane) first came here,' says Armstrong, 'there was great scarcity of fresh provisions: Goat's-Flesh indeed might be had; but there was little beef or mutton, and tame fowls were a greater rarity than the wild. Mr Kane procured numbers of Cattle and Flocks of sheep; he had supplies of Poultry from France, Italy and Barbary, and distributed them, together with great quantities of eggs, among the Farmers and Peasants'.[23] Kane introduced clover (still known as *enclover*) from England with the object of assisting the cattle industry; he also began the cultivation of several English fruits and vegetables, and one brand of apples is still known after him, *Pomes d'en Ken*. In 1726 and again in 1732 farmers suffered very heavy losses of stock. On the first occasion the loss was due to an unusual snowstorm, but in 1732 over 2,000 cattle, 10,000 sheep, 1,200 goats and 1,100 pigs died owing to shortage of fodder. Thereupon Kane gave orders that cattle and sheep be imported from Barbary and pigs from Sardinia.[24] He fixed the prices of meat and fish, and gave orders that licences must be issued to all persons who sold bread. Another decree laid down that farmers must keep a statistical record of their harvest of cereals. By an order issued in 1728, Kane enacted that farmers must not reap their corn until it was ripe, and also that they must cut the corn 6in from the ground, and keep the straw in a dry place.[25] Hares from England were introduced by him, but the British devotees of hunting soon exterminated them.[26] Kane issued a strict decree about the sale and price of wine. No new wine was to be sold to English soldiers before December 21, and the price of red wine was fixed annually. The consumption of wine was considerable; it was estimated that 13,000 hogsheads of wine were produced annually in Menorca and that 10,000 hogsheads were sold to the English for £17,500. Of the remainder, 2,000 hogsheads went to the clergy and 1,000 to the islanders.[27] Cleghorn believed that the drinking of new wine was the cause of the dysenteries so prevalent among the English, and both he and Armstrong remark on the excess of drinking that was a universal vice of the troops.[28]

In times of peace, five British 'regiments' (battalions) were quartered in Menorca. One was stationed at Mahón, another at Alayor with a detachment at Fornells, and another at Ciudadela. The others were quartered in the Fort of St Philip with a company of artillery and a number of engineers. Ciudadela seems to have been the favourite station of the British, as the best quarters were found there.[29]

The exports of Menorca were of no account and the island was obliged to import one-third of the corn required, and all the oil and spirits. The excess of imports over exports was balanced by the English money circulated by the troops and used to purchase provisions. It was estimated that the production of vineyards, vegetables and the breeding of fowls was increased about fivefold in the first forty years of the British occupation.

Kane endeavoured to control the price of foodstuffs by fixing tariffs, but increased demand often sent up prices, for 'when a fleet of ships lies in Port Mahón the price of fowls rises with the demand, and is sometimes more than doubled, but they are no sooner gone than the markets return to the old standard again'.[30]

During the British occupation more than twice as much wheat as barley was grown in the island in a normal year. In 1752 the island contained between 6,000 and 7,000 cattle, 60,000 sheep, 20,000 goats and 4,000 swine. The total food resources were insufficient for the needs of the troops and they were supplemented by cheese imported from England, and by butter, salt beef, neats' tongues and potatoes, which were all sent from Ireland. Clothes, malt-liquor, cider, books and other luxuries came from London. The white wine of Bañalbufar in Mallorca, the best wine produced in the Balearics, was imported for the use of British officers.

Kane died in Menorca on 29 December 1736. Armstrong writes of him: 'The gentleness of his administration reconciled the Minorquins to the English Government; and the Troops observed an exact discipline under so nice a judge.'[31] Kane's policy of benevolent government was continued by his successors until, in 1756, the first British occupation came to an end, when the island was captured by the French. In April of that year a French army of 15,000 men under the Duc de Richelieu landed at Ciudadela, marched across the island and surrounded Fort St Philip. The British had broken up Kane's road near Mercadal, and it would have been difficult to move the artillery across the island. For this reason the French landed their siege train at Cala Mesquida. Admiral Byng failed to relieve Menorca, as he was beaten off by a French fleet on May 20 and retired to Gibraltar. The Fort was gallantly defended by a garrison of about 3,000 under the command of General William Blakeney, the Lieutenant-Governor, a veteran of 84, who was compelled to capitulate on June 29. Blakeney was given a peerage; Byng was executed.

The French occupation of the island lasted from 1756 to 1763, and was marked by the foundation of a new town called San Luis. This place was carefully planned with four main streets running north to south, and five at right angles to them; its streets still bear French names. The town was built with the object of creating a nucleus between the dispersed barracks of that region. The French also built a road, 222 kilometres in length, the whole way round the cliffs of the island.[32] Batteries were placed at various points round the island, and this circular road was constructed in order that troops might be rushed to any threatened part of the coast. This road has, however, long ceased to be of any importance. In addition to this, the French built a further 90 kilometres of military roads.

The island was restored to Britain in 1763 by the Treaty of Paris, and remained British for eighteen years. The second British occupation was not so paternal as the first, but a considerable amount of building was carried out. Under General Moystin a new town, called George Town (now called Villa Carlos), was built. It was realised that the old settlement which had grown up immediately outside the walls of the fort was inconveniently close. The old houses were destroyed, and an entirely new town with barracks for two battalions was built on the edge of the harbour, about two miles east of Mahón. New barracks were built in Mahón in 1765, and a new naval hospital in 1771–6 to replace the old one on Bloody Island (Plate 6).

In 1781 a combined force of Spanish and French troops, numbering about 16,000 under the Duc de Crillon, attacked the island. They landed on 19 August, partly in the Cala Mesquida and partly in Cala Alcaufa. St Philip's Fort, defended by 2,200 men under General Murray, capitulated in January 1782. After sixteen years of Spanish rule the island fell into the hands of the British for the third time in 1798. In November, a force of about 4,000 men, under General Charles Stuart, landed on the north of the island at Addaya and seized Mercadal. The Spanish main force defended Ciudadela, so Mahón was easily captured and Ciudadela surrendered after an amazing piece of bluff by Stuart; there was not a single British casualty. The third British occupation lasted for four years only, and although the times were mainly devoted to military affairs, Stuart's rule was benevolent, and he was able to undertake several reforms in the municipal government of the island. In the year 1800, no less than 18,000 British troops were quartered in Menorca. By the Treaty of Amiens, however, the island was returned to Spain in June 1802 and has ever since been in Spanish possession.[33]

During the Peninsular War, Menorca was protected by British squadrons, and the continuous naval activity gave a prosperity to the island. As soon as the Napoleonic Wars were over, Menorca ceased to flourish. The introduction of steam was a further blow

73

to the shallow harbour of Mahón, and the early nineteenth century was a period of decay and poverty. Between 1835 and 1840 several thousands of Menorcans left their island, partly to escape the hated conscription. Many of them found a new home in Algeria, where they became admirable colonists, in the region which had just been taken over by the French. Menorcans settled in Oran at Bel-Abbès, and gave Algeria the benefits of their farming experience in an arid land. The comparative proximity of Algeria, and the consequent easy communication with home combined with a climate similar to that of Menorca, made it a very suitable region for Menorcan emigrants.

The British occupation was then regarded as the most prosperous period in the island's history, and as late as 1876, Mr C. T. Bidwell, the British Consul at Palma, wrote that some of the inhabitants of Menorca 'express in broken English their love for England, while they speak joyously and feelingly of the good and flourishing times when Minorca was under British rule' and also stated that he often heard 'the period of the British dominion spoken of as the really good times which, it is regretted, have disappeared for ever.[34]

The town of Mahón still bears the marks of its connection with England. The small, clean houses, with lace blinds, green doors, sash-windows with square panes, an occasional bow-window, and the white woodwork remind one forcibly of an English town, and appear very strange in the Mediterranean world. (Plate 4 (*a*).) The English type of sash-window is found in all parts of the island, and is considered to be more satisfactory as a protection against wind than the usual kind found in this part of Europe. The town of Ciudadela, on the other hand, is markedly Spanish in its architecture and general appearance. Its narrow streets, some flanked with arcades, and its small squares are in marked contrast to the plan and character of Mahón (Plate 4 (*b*)).

The number of English words still in use, most of them in a corrupt form, is over one hundred: many of them are concerned with drink, carpentry and games.[35] A few Christian names are still in use in the English form instead of the Spanish, as for example, Piter, Jims, Tony, Fanny, Jo, Jery, Eliza, Jony. When much noise is going on in a house, the expression, *semble un bèrious* is used, and can be translated 'It is like a barrack house!' A certain brand of plum is known as *prunes de nèversó*. The origin of this name is believed to be connected with Governor Kane, who visited the market every day. One one occasion, a woman in the market showed Kane a plum and asked him what this variety was called in England. The Governor replied 'I *never saw* them before,'[36] and from that time they were known in Menorca as *prunes de nèversó*. The drive from Mahón to Villa Carlos, San Luis and San Clemente, a distance of nearly ten miles, is known as the *vuelta de Milord*, as it is said to have been a favourite walk of the British.

Neither the number nor the distribution of the population of Menorca showed any very marked change between the close of the British occupation and 1930. The total population, which numbered 28,000 in 1756 and was estimated at 32,000 in 1800, has slowly increased, though the stream of emigration to Algeria and later to the Argentine, Uruguay and Cuba, checked its natural growth. In 1860, there were 37,000 inhabitants of Menorca; in 1900, the same; in 1911, 39,000; in 1920, 42,000; and in 1930, 41,500. A map of the distribution of population in 1930 showed that the major settlements were still the two ports of Mahón (16,000) and Ciudadela (8,000). The inland villages are all placed near the junction of the two geological regions, and enjoy a certain amount of shelter from the winds, as they lie in the lee of the northern hills. The three largest inland nuclei are Alayor (4,300), Ferrerias (1,030), and Mercadal (1,000). The last named is protected by Monte Toro, and the other two villages are also sheltered by hills. It is noticeable that dispersed settlement is somewhat denser on the southern side of the plateau than on the northern.

The most important fact shown by a map of the distribution of population in the island is the predominance of Mahón and its surrounding region. This characteristic is a legacy of the British occupation. The capital and its satellite towns of Villa Carlos (2,000). San Luis (890), and villages like San Clemente (280) together contain more than half the population of the island. Thus, the region within a few miles of the harbour is the region of greatest density of population in Menorca. Ciudadela contains less than one-fifth of the island's population and has no outlying suburbs. When the British first occupied Menorca, Ciudadela and its region was of greater importance than the Mahón region. The removal of the island's centre of gravity from Ciudadela to Mahón was brought about for military reasons. Some historical geographers deny the importance of military affairs in the study of geography. But, if geography includes the study of human settlement, how can the growth of towns like Aldershot, Portsmouth and Mahón be explained without reference to military and strategic considerations?

NOTES

1 This chapter was read to Section E (Geography) of the British Association for the Advancement of Science, at Norwich on 6 September 1935, and appeared in *Scott G Mag* 52 (1936), 375–90, as 'Influences of the British occupation on the geography of Menorca'. The present version of this paper owes much to suggestions and corrections made by P. Dennis of Halesworth. No attempt has been made to bring agricultural and population statistics beyond the year 1930.

2 Hermite, H., *Études Géologiques sur les Îles Baléares* (1879).

3 Vila, P., 'Le clima de Minorque', *Rev de Géog Alpine*, 21 (1933), 831–9

4 Guardiola, J. M. Jansá, 'Contribución al estudio de la Tramontana en Menorca', *Memorias del Servicio Meteorológico Espanol*, Serie A.3 (1933), and *Régimen de Vientos* (Mahón, 1934).

5 Knocke, H., *Flora Balearica: Étude Phytogéographique dur le Îles Baléares*, 4 vols (1921–3). This work contains maps of the geology and vegetation of Menorca.

6 Armstrong, J., *The history of the Island of Menorca* (1752), 256.

7 ibid, 3.

8 Cleghorn, G., *Observations on the Epidemical Diseases in Minorca from the year 1744 to 1749, to which is prefixed a short account of the*

climate, production, inhabitants and endemical distempers of the Island (1751), 45.

9 *Memoria Comercial de la Cámara Oficial de Comercio, Industria y Navegación de Menorca* (1934), 2.

10 Armstrong, J., op cit, 213

11 ibid, 95–6

12 Campbell, C., *The Ancient and Modern History of the Balearick Islands* (1716), 304.

13 Victory, Antonio, *Gobierno de Sir Richard Kane en Menorca (1712–1736)* (Mahón, 1924).

14 Armstrong, J., op cit, 97

15 ibid, 253

16 ibid, 63

17 ibid, 63

18 PRO, Cal Treas Papers 1708–14, 179, 41; and Brit Mus, Add MSS 17775.

19 BM, Add 23638, f 38/39.

20 Sanz, F. Hernández, *Revista de Menorca*, 29 (1934), 168–9.

21 Tunstall, B., *Admiral Byng and the Loss of Minorca* (1928), 93.

22 Armstrong, J., op cit, 263.

23 ibid, 25.

24 Victory, Antonio, op cit, 55.

25 *Libro de ordenes de Gouvernador (Richt.) Kane*, fol 26. Documentación de la Real Gobernación de Menorca.

26 Armstrong, J., op cit, 155.

27 ibid, 126–7. See Bisson, Jean, *La Tierra y el Hombre en Menorca* (Palma, 1967), for a map of natural and cultivated vegetation in the island in 1770 during the British occupation.

28 ibid, 251; also Cleghorn, G., op cit, 73.

29 ibid (Armstrong), 66

30 ibid, 245.

31 ibid, 25.

32 *A Topographical Map of the Isle of Minorca geometrically survey'd by the Royal Engineers while it remained in the possession of the French during the last War and digested by L. S. de la Rochette* (London, 1780).

33 For a more detailed account of the political and military history of Menorca in the eighteenth century, see Sanz, Francisco Hernández, *Compendio de Geografía e Historia de la Isla de Menorca* (Mahón, 1908).

34 Bidwell, C. T., *The Balearic Islands* (London, 1876), 307–8.

35 The following is a list of some of the English words and phrases in use in Menorca in 1915: recess, stirrup, boy, bottle, mug, punch, grog, rum, gin, peppermint, shoe-maker, scour (verb), strop,

halloo, chalk, rule, cap, blackball, black-varnish, screw, turnscrew, plenty, ox, sideboard, teaboard, white, shank, pilchard, mitre, gravy, bow-window, mahogany, panel, settee, squeak, floor, back, bit, rail, lath, midshipman, 'shake hands', razor, pudding, clog, panel, bargain, bargainer, kettle, taper, batten, rabble, pinch, lazy, beggar, taste (verb), haversack, bitter, jack (tool), barracks, barrack-house, brad, gaiters, up-up (admonition to children), bully, can, shell, jacket, gang, penny, 'to shake off', black, beef, fight. The following words were used in games: stick (billiard cue), 'even all', kiss, pink, in, out, play, stop (in bowls), marbles, please (in bowls). See Escudero, Bartolomé, 'Lista de varias de las palabras que usamos en Menorca tomadas del Inglés durante las tres dominaciones Británicas de la Isla,' *Revista de Menorca*, 10, 169–74, 222–3, 257–9 (Mahón, 1915).

36 Victory, A., op cit, 54

CHAPTER FOUR

Victorian Pioneers in Medical Geography

A**T THE PRESENT** time there is a lively interest in medical geography. A great atlas of the distribution of diseases is being compiled under the auspices of the American Geographical Society. The International Geographical Union has established a Commission on medical geography, defined as 'the study of geographical factors concerned with cause and effect in health and disease:' the Commission presented its first report at Washington in 1952. British geographers have published papers on the ecology of disease in such journals as the *Practitioner* and the *British Medical Journal*. In view of all this activity it is right that the work of the nineteenth-century pioneers in the drawing of maps of health and disease should be remembered, especially as some of the leading medical cartographers were Fellows of the Royal Geographical Society. The object of this chapter is to describe a few of the maps constructed by these pioneers and to explain the significance of their cartographic achievements.[1]

Much of the subject matter of what is now called medical geography is as old as Hippocrates and his *De Aëre, Aquis et Locis*. Even the term 'medical geography' is not new[2]. It has been current in England for at least eighty years, as it was used by Dr Alfred Haviland in his *Geographical Distribution of Diseases in Great Britain*, published in 1892. In the introduction to this book Haviland wrote of 'medical geography', and on p 286 he referred to 'the medical geography of cancer'. In 1891 he read a paper in London to the Seventh International Congress of Hygiene on 'the influence of clays and limestones on medical geography; illustrated by the geographical distribution of cancer among females in England and Wales'. But it appears that the term 'medical geography' was preceded by 'medical topography'. For instance, in 1830 Dr John Hennen published his *Sketches of the Medical Topography of the Mediterranean*, which contains full and systematic geographical accounts of Gibraltar, Malta, and the Ionian Islands. In this book the geography of the areas named, all of which were then occupied by British naval and military forces, is described in relation to health and disease. The author also prepared an elaborate scheme for publishing a series of 'Memoirs on medical topography', but his plan was never carried out. He proposed that each memoir should describe the physical geography including the relief, climate and vegetation of selected areas, and that this description should be followed by an account of the population, its dwellings, bedding, clothing, furniture, diet, employment, amusements, customs, morals and so forth. A third part of each memoir was to be devoted to diseases, hospitals, longevity and vital statistics. His own book contains material of this kind and is a valuable source of information about conditions then prevailing in Mediterranean lands.[3]

Books on medical topography appeared in many countries in the nineteenth century. Some of these dealt with single towns, for example, M. Murat, *Topographie Médicale de la Ville de Montpellier* (1810); and R. H. Powell, *A Medical Topography of Tunbridge Wells* (1846). Other books described provinces or districts: J. B. Pesétri y Vidal, *Topografía Médica de Valencia* (1878) was one of several similar works on the medical topography of Spanish provinces. In England, partly as a result of the efforts of Dr (later Sir) James Clark (1788–1870), whose well-known book on *The Influence of Climate in the Prevention and Cure of Chronic Diseases* was first published in 1829,[4] a considerable number of works on what might be called the 'medical climatology' of English towns and districts began to appear. Notable among these was a detailed memoir on the medical topography of a part of Cornwall by Dr John Forbes (1787–1861), a schoolfellow and life-long friend of Sir James Clark.[5] Other contributions were Dr Thomas Shapter's *The*

Climate of the South of Devonshire; and its Influence upon Health (1842); and Dr A. L. Wigan's *Brighton and its Three Climates with Remarks on its Medical Topography* (1843). In his book on climate Dr Clark praised the then unknown Ventnor in the Isle of Wight as a health-giving winter resort; his commendations sent up the price of building land at that seaside village from £100 per acre to £800 or £1,000 in a very few years.

These early writers on 'medical topography' did not include the mapping of the distribution of diseases as part of their studies. The great outbreaks of cholera in the first half of the nineteenth century seem to have been the factor which first stimulated carto-graphic work of this kind. It has been stated that the period 1835–55 can be regarded as 'a golden age' of the development of geographic cartography',[6] 'and it is true to say that the mapping of diseases began in England during those years as did the map-ping of the distribution of population.

Asiatic cholera reached Great Britain in October 1831 the first death occurring in that month at Sunderland.[7] During the epi-demic of 1831–2 over 52,000 persons died of cholera in the British Isles, including nearly 22,000 in England and Wales, 21,000 in Ireland and over 9,600 in Scotland. In 1848–9 a second epidemic of cholera caused over 53,000 deaths in England and Wales alone; while in 1853–4 there were over 20,000 deaths from this cause in the two countries; and over 14,000 in 1866. There were a few cholera deaths in 1873, and in 1893 there were 135 deaths, this being the last year in which cholera gained a foothold in Britain.

It is well known that Dr John Snow, (1813–58, Plate 9) a famous London anaesthetist, was largely responsible for demonstrating the water-borne origin of cholera. In his short essay *On the Mode of Communication of Cholera*, published in 1849, he urged the necessity of providing sewage-free water for south and east London as a preventive of cholera: the second edition (1855) was a substantial enlargement of the original pamphlet.[8] Moreover, the second edition included a map of the distribution of deaths from cholera in the Broad Street district of London in 1854. The 'cholera field', as Snow called it, had its centre at a pump in Broad Street near Golden Square, and the 'field' was bounded roughly by Great

Marlborough Street, Dean Street, Brewer Street and King Street. In this small space in Soho there were over 500 deaths in ten days of September 1854. The scale of the original map is 30 in to 1 mile; deaths are shown by black rectangles and the pumps are also marked (Fig 4). Snow proved 'that the incidence of cholera was only among persons who drank from the Broad Street pump'. On 8 September, at Snow's urgent request, the handle of the Broad Street pump was removed and the incidence of new cases in the

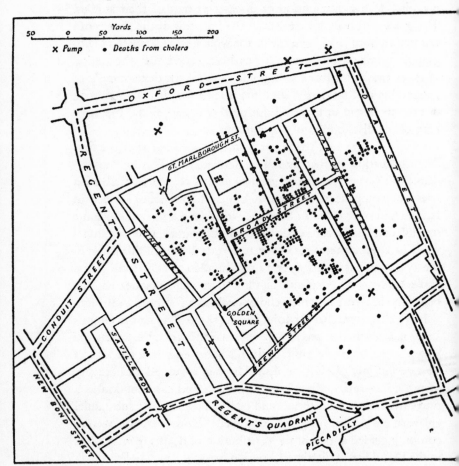

Fig 4   Dr John Snow's plan (1855) of deaths from cholera in the Broad Street area (Soho) of London in September 1854

area ceased almost at once. Snow's map of the distribution of deaths from cholera, first published in 1855, is a very significant document in the history of medical geography, but it is not the first map of its kind.

The 1832 outbreak of cholera was particularly severe in York-shire.[9] In Leeds, then a place of 76,000 people, there were over 1,800 cases and 700 deaths between May and November of 1832. Dr Robert Baker illustrated his *Report of the Leeds Board of Health* (1833) with a 'cholera plan', surveyed by Charles Fowler, on which the parts of Leeds affected by cholera were marked and coloured red (Fig 5). Dr Baker's map showed that the incidence of the disease was highest in the more densely peopled and north-eastern districts of the town and that comparatively small areas were

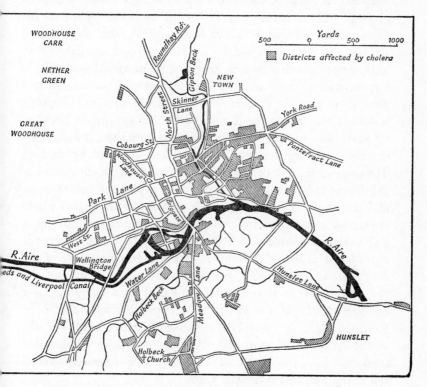

Fig 5   Dr Robert Baker's 'cholera plan' of Leeds (1833)

affected in those parts of Leeds that lie south of the river Aire. The plan is on a scale of about 3½in to 1 mile and indicates by hatching 'the districts in which the cholera prevailed'; it does not mark individual cases or deaths (Fig 5). Dr Baker observed (p 10) 'how exceedingly the disease has prevailed in those parts of the town where there is a deficiency, often an entire want of sewerage, drainage, and paving'.

In 1848 W. P. Ormerod in *The Sanatory Condition of Oxford* included a rather crude plan of the city on the scale of about 12in to 1 mile to show the districts visited by cholera and by fever in 1832. Localities of fever were marked by a cross and those of cholera by a dot, while the parts of the city 'chiefly visited by disease generally' were slightly shaded. In 1849 Dr Thomas Shapter (1809–1902), who had a reputation for his work in both geology and climatology as well as in medicine, published his *History of the Cholera in Exeter in 1832*. This admirably written account has been described by Dr E. Ashworth Underwood as 'one of the best descriptions extant of an historical epidemic'.[10] Exeter, which then contained a population of about 28,000, had over 1,100 cases of cholera with 402 deaths in 1832. Shapter's book is important in the history of medical geography because, six years before Snow's famous map was published, it included a dot map showing the distribution of deaths caused by cholera. The original map, drawn on a scale of 6in to 1 mile, distinguishes the deaths which took place in the years 1832, 1833 and 1834, by different symbols for each year. It also marks the places where clothes were destroyed, the cholera burying grounds, the druggists and the soup kitchens. A redrawing of this plan of Exeter is reproduced as Fig 6, and shows that a large proportion of the deaths occurred in the low-lying south-eastern quarter of the old walled city. In 1853 Shapter published his *Sanitary Measures and their Results*, as a sequel to his previous work on cholera in Exeter in 1832. He pointed out that Exeter suffered only 43 deaths in the epidemic of 1849 and attributed the city's comparative immunity from the disease to the sanitary improvements carried out in the previous seventeen years.

In 1852, Augustus Petermann (1822–78), the distinguished

Plate 5   Bust of Governor Richard Kane by M. Rysbrack in Westminster Abbey

Plate 6   Mahon harbour from a map of Menorca 'survey'd by the Royal Engineers while it remained in the possession of the French during the last war, and digested by L. S. de la Rochette 1780.' Engraved by William

German geographer, at that time resident in England, brought out a *Cholera Map of the British Isles Showing the Districts Attacked in 1831, 1832 and 1833*. A redrawn copy of this map is reproduced as Fig 7.[11] Petermann had worked with the firm of Berghaus at Potsdam from 1839 to 1845. He then migrated to Britain, first living in Edinburgh from 1845 to 1847, where he worked with Alexander Keith Johnston; and then in London until 1854.[12] He was elected a Fellow of the Royal Geographical Society in 1846 and in 1852 was described as 'Physical Geographer and Engraver on Stone to the Queen'. His cholera map, drawn on a scale of 1: 2,200,000 shows 'all districts visited by the cholera' by means of crude shading, 'which is darker in proportion to the relative amount of mortality'. The map is accompanied by a few pages of 'Statistical Notes' which are of considerable geographical interest. Petermann's introduction to these notes is worth quoting in full:

While the means of intercourse between distant nations were in their infancy, the knowledge of the existence or development of remarkable diseases was necessarily limited, and consequently that department of science, which may be termed the *Geography of Diseases*, remained uncultivated. Now, however, as we live in a different era, we may indulge in the rational hope of seeing this science studied which, though comparatively new, is one of the most important branches of Geography.

The object, therefore, in constructing Cholera Maps is to obtain a view of the Geographical extent of the ravages of this disease, and to discover the local conditions that might influence its progress and its degree of fatality.

For such a purpose, Geographical delineation is of the utmost value, and even indispensable; for while the symbols of the masses of statistical data in figures, however clearly they might be arranged in Systematic Tables, present but a uniform appearance, the same data, embodied in a Map, will convey at once, the relative bearing and proportion of the single data together with their position, extent, and distance, and thus, a Map will make visible to the eye the

F

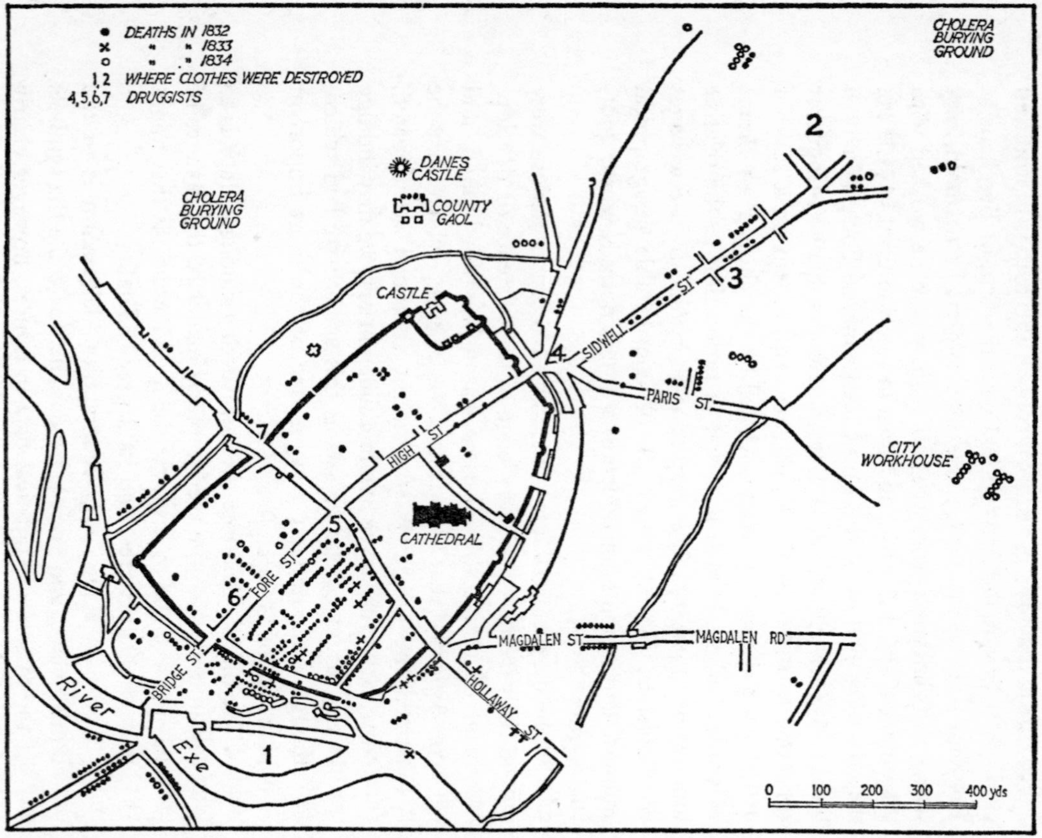

DEATHS IN 1832
" " 1833
" " 1834
1,2 WHERE CLOTHES WERE DESTROYED
4,5,6,7 DRUGGISTS
CHOLERA BURYING GROUND
DANES CASTLE
COUNTY GAOL
CHOLERA BURYING GROUND
CASTLE
SIDWELL ST
PARIS ST
HIGH ST
CATHEDRAL
CITY WORKHOUSE
FORE ST
BRIDGE ST
MAGDALEN ST
MAGDALEN RD
HOLLAWAY ST
River Exe
0 100 200 300 400 yds

Fig 6 Dr Thomas Shapter's plan (1849) of the deaths from cholera in Exeter in 1832–4

development and nature of any phenomenon in regard to its geographical distribution.

Petermann then described the spread of the disease through the country in 1832. He pointed out that it travelled slowly from Sunderland along the inland lines of communication, but 'more rapidly to the different seaports, so that all the coast line round the Isles was first attacked, and from thence the disease penetrated the inland parts of the country'. He also discussed the relation of the disease to relief and observed that a cursory glance at his map apparently corroborated the current belief that cholera rarely penetrated mountainous countries and never reached the tops of hills. The cholera districts in the British Isles 'seemed to lie all in the lower ground and valleys'. When he began his geographical investigations Petermann was predisposed in favour of this general opinion, 'and if such had proved to be the case, it would have been of great interest to ascertain the scale of perpendicular distribution of cholera, the relative amount of mortality in the different stages of elevation, and the point where (like the eternal snow limit which throws a bar against animal and vegetable life at large) the disease would altogether cease. But, by a more minute investigation, it appears with a considerable degree of certainty that these districts were attacked, not so much in conse-quence of their *low situation*, as from the *great amount of population they contain*'.

In 1848 Petermann had compiled and drawn a fine population map of the British Isles on the basis of the census of 1841.[13] He was later called on to design a population map of England and Wales and another of Scotland to illustrate the 'Census of Great Britain, 1851' (1852). These two engraved maps, drawn on a scale of $31\frac{1}{4}$ miles to 1in, show, by very delicate shading, the differences in density of population, the darkest shading being used for a den-sity of 600 persons and upwards per square mile. Petermann said in his 'Statistical Notes' that a comparison of his population maps with his cholera map showed 'that the more densely peopled dis-tricts were proportionately the most severely attacked'. He said that while the 'great level of the Fens' did not suffer much from cholera, 'the elevated land of Birmingham, forming a pretty ex-

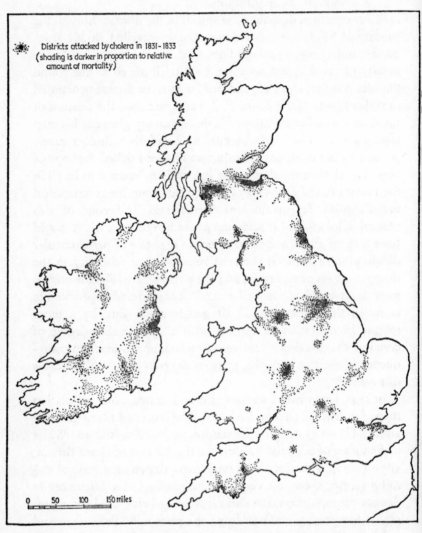

Districts attacked by cholera in 1831-1833
(shading is darker in proportion to relative
amount of mortality)

0   50   100   150 miles

*Fig 7   Augustus Petermann's cholera map (1852) of the British Isles
showing the areas affected in 1831-3*

tensive plateau of 500 feet mean elevation' was one of the most severely afflicted parts of the whole kingdom. This proved, according to Petermann, that altitude was of little significance in the spread of cholera and that density of population was of far greater importance. He also included some interesting remarks on the influence of the seasons on the progress of the disease, illustrated on his map by an inset diagram of 'the number of places of Great Britain attacked by the disease in different months'. As a result of his investigations he concluded that 'the warmer season favours the progress of the disease,' and that 'cholera forms a tolerably regular course, ascending rapidly from May to August, attaining in that month its summit, and then descending again with equal rapidity till November, from which month till May it keeps a pretty low ebb'.

Petermann's cholera map of the British Isles also contains an inset plan of London divided into 43 registration districts; and detailed tables of the number of cases of cholera, the deaths, the total population and the proportion of deaths to population in each district. On this plan (Fig 8) six different tints of red or pink were used to indicate the varying proportions of deaths from cholera to total population in the outbreak of 1832. Petermann adopted a scale ranging from one death in 35 persons for the worst district of Botolph (without Aldgate) to one in 900 persons or more for areas least affected and applied the appropriate tint to each registration district. The plan gives some idea of the distribution of the districts most seriously affected in 1832; they included Southwark, Bermondsey and the City.

About 5,000 persons died in London in the outbreak of 1832, another 1,000 in 1848–9, and over 10,700 in the severe epidemic of 1854.[14] In 1866 there were over 5,500 deaths in London, mainly distributed in the East End and not so largely in Southwark and other parishes south of the Thames as in previous outbreaks.

In 1856, one year after the publication of John Snow's map of cholera in the Broad Street area of London, Dr Henry Wentworth Acland (1815–1900, Plate 10) produced his *Memoir on the Cholera at Oxford in the Year 1854*. Oxford experienced three outbreaks of cholera, with 95 deaths in 1832, 75 in 1849 and 129 in 1854.

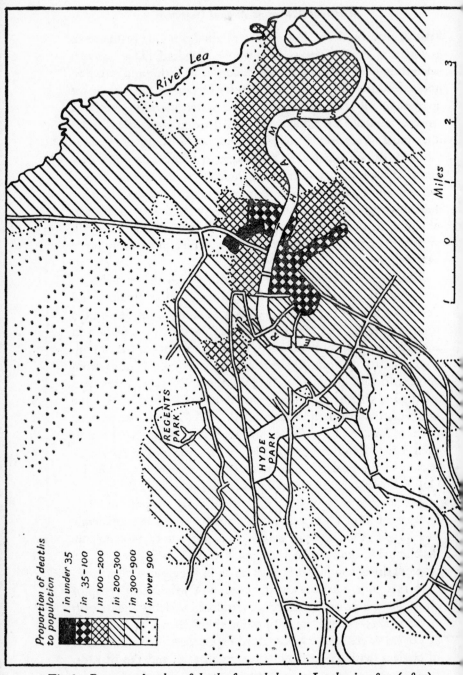

Fig 8   Petermann's plan of deaths from cholera in London in 1832 (1852)

Like Petermann, Dr Acland was a Fellow of the Royal Geographical Society, having been elected in 1855, and his geographical interests are clearly revealed by the remarkable maps in his book. His detailed plan of the city of Oxford is on a scale of approximately 10in to 1 mile. In addition to marking rivers and streams the plan has contours drawn at intervals of 5ft below the summit of Carfax, the highest point in the old city (Fig 9). Carfax is marked as 49·64ft above an assumed datum which is a point 12ft below the base of the monument at Sandford Weir. By this curious arrangement the first 5ft contour below Carfax is marked 44·64ft, the next 39·64ft and so on. Contours had been

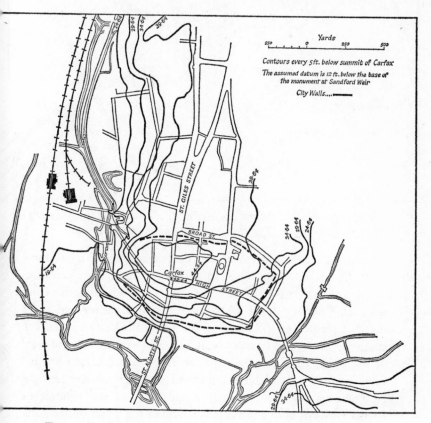

Fig 9   Dr H. W. Acland's contoured plan of Oxford (1856)

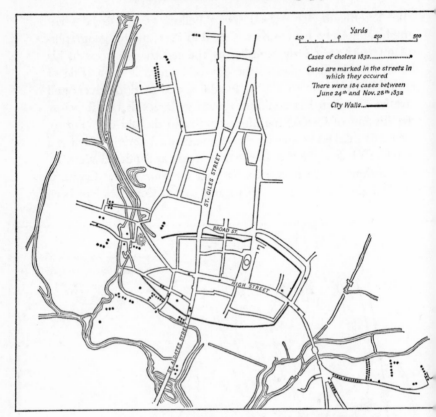

Fig 10   Dr H. W. Acland's plan of cases of cholera in Oxford in 1832 (1856). In 1832 there were 184 cases and 95 deaths

introduced into the maps of the Ordnance Survey by Colonel T. F. Colby, RE (1784–1852) in about 1830, but it was still un-usual in 1856 to make a plan of a city with contours at such a close interval. Certainly these 5ft contours give a clearer and more detailed picture of the relief of the old walled city of Oxford, standing on a gravel terrace about 25–30ft above the normal level of the river, than do most modern maps. Dr Acland obtained these contours from the unpublished drawings of Robert Syer Hoggar, an engineer. Hoggar calculated that the summit of Carfax was 37ft above the average level of the water under Folly Bridge. The difference between Carfax (209ft above sea level) and

94

the average level of water under Folly Bridge (179ft above sea level) was 30ft. The difference of 7ft between the measurement of 1856 and that of 1958 is mainly due to changes in the level of the river which occurred after the construction of weirs and partly to alterations at Carfax itself. Hoggar's published plan of Oxford is dated 1850, but the copy in the Bodleian shows contours in manuscript only. The plan shows by colour and symbols the parts of the city still undrained in 1855, and the rivers and streams that were then contaminated by sewers as well as the points of contamination (Fig 13). By blue symbols of different shapes the cases of cholera in the 1832 epidemic can be differentiated from those of

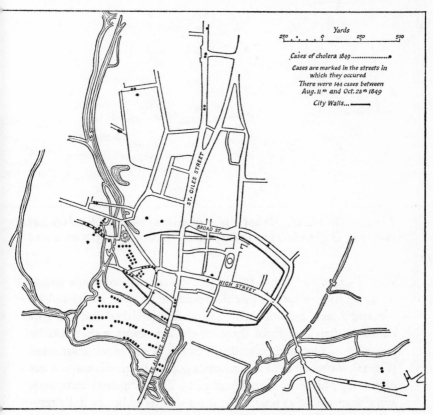

Fig 11   Dr H. W. Acland's plan of cases of cholera in Oxford in 1849 (1856). In 1849 there were 144 cases and 75 deaths

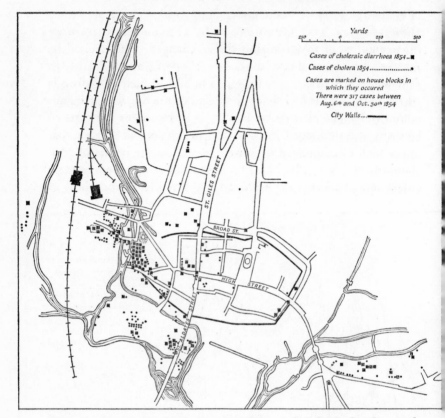

Fig 12   Dr H. W. Acland's plan of cases of choleraic diarrhoea and cholera in Oxford in 1854 (1856). In 1854 there were 317 cases and 129 deaths

1849 (Figs 10 and 11); both sets of cases are marked in the streets in which they occurred. For the epidemic of 1854, by using two different black symbols, Dr Acland distinguished the cases of choleraic diarrhoea from those of cholera, and for that year he was able to plot each case in the actual house block concerned (Fig 12). Acland's map also marks by special symbols those parts of the city which had been described by W. P. Ormerod and others some years earlier as unhealthy; those which had been wholly remedied; and those which had been partly remedied since they wrote.

Acland's plan of Oxford is the most elaborate in the whole series of cholera maps. It is not possible to discuss Acland's text fully, but it should be noted that, like Petermann, he carefully considered the relation of cholera mortality to altitude. Acland did not reach the same conclusions as Petermann, but followed the opinions expressed in the Registrar General's Report on cholera in 1849, that altitude in London had 'a more constant relation with mortality from cholera than any other human element'. Acland took his contour line marked 29·64ft (Fig 9) as a division between an upper and a lower level in Oxford. This contour was 16·47ft above the average water level at Folly Bridge and 20ft below Carfax. He calculated that in the three Oxford epidemics 141

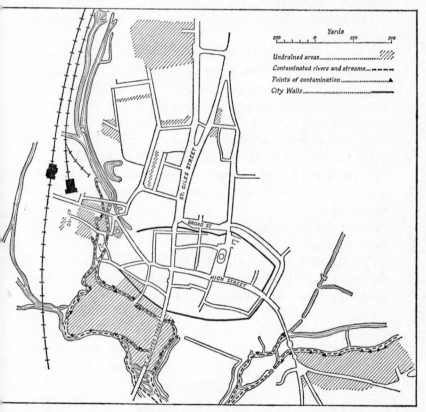

*Fig 13   Dr H. W. Acland's map of undrained districts in Oxford
1854 (1856)*

cases occurred in the upper level and 362 in the lower, and estimated the population in the upper level at 14,200, and that in the
lower at 12,300. He concluded 'that mortality on our lower level
was proportionately three times as great as that of our upper
level'.[15] Acland's contour correlations are explained by drainage,
as at lower levels the opportunities for contamination were much
greater. Acland also gave a full description of the meteorological
conditions during the outbreaks, with elaborate diagrams, as part
of his attempt to correlate weather with disease. 'The publication
of the cholera memoir', said J. B. Atlay, 'raised Acland from the
position of a country physician in good practice to that of an
authority on sanitary and hygienic questions, of acknowledged
repute not only at home, but on the Continent, and in Canada and
the United States'.[16]

A brief reference should be made to two pioneer health maps.
In 1882 Dr Alfred Haviland produced a *Health-Guide map for
Brighton* and in the following year a similar map for Scarborough.[17]
The Brighton map, drawn on a scale of 9in to 1 mile, showed
geology in colour and relief by contours at 25ft intervals. The
most unusual feature of the map was the indication of 'aspect'
which was shown by arrows and letters. Arrows point in the
direction of the slope and letters show the wind point to which
the slope was exposed. Wind charts and climatic statistics fill the
borders of the map. The object of these two maps was to indicate
the varying merits of different parts of the towns concerned in the
matter of health.

The Victorian doctors who brought cartography to the aid of
medicine were indeed pioneers in geography and their work gave
immeasurable benefit to Britain and the world.

NOTES

1 An outline of this chapter was read as a paper at the Washington
   meeting of the IGU, to the Commission on Medical Geography on
   11 August 1952, on behalf of the author by Professor R. W. Steel.

The paper was printed in *Geogr J*, 124 (1958), 172–83 as 'Pioneer maps of health and disease in England'.

2 In Germany the term 'medical geography' was used earlier than in England. A book entitled *Medizinische Geographie*, by Fuchs, C. F., was published at Berlin in 1853. For the history of maps of disease in Germany see Jusatz, H. J., 'Zur Entwicklungsgeschichte der medizinisch-geographischen Karten in Deutschland', *Mitt Reichsamts Landesaufn*, 1939, 11–22.

3 An appendix on 'Medical Topography of the different military stations in Jamaica' appeared in Arnold, Dr W., *Practical Treatise on the Bilious Remittent Fever* (1840).

4 Sir James Clark (as he had become) republished this book in 1841 under a new title: *The Sanative Influence of Climate: with an account of the best places of resort for invalids in England, the south of Europe, etc.* As a young man Clark had assisted W. E. Parry, the Arctic explorer, to make some experiments on the temperature of the Gulf Stream.

5 Forbes, John, 'Sketch of the medical topography of the Hundred of Penwith, comprising the District of Landsend in Cornwall,' *Trans Provincial, Medical and Surgical Assoc*, 2 (1834), 32–147; 4 (1836), 152–261. One of the aims of this Association was to promote the 'increase of knowledge of the medical topography of England, through statistical, meteorological, geological and botanical enquiries' and its early volumes include a number of papers on medical topography.

6 Robinson, Arthur H., 'The 1837 maps of Henry Drury Harness', *Geogr J*, 121 (1955), 440.

7 Underwood, E. Ashworth, 'The history of Cholera in Great Britain', *Proc R Soc Med*, 41 (1948), 165–73.

8 Snow's important papers have been reprinted in *Snow on Cholera, being a Reprint of two Papers by John Snow, M.D.* (New York, 1936), with a memoir by Richardson, B. W., and an introduction by Frost, W. H. The Broad Street map is reproduced in this book.

9 Underwood, E. Ashworth, 'The history of the 1832 cholera epidemic in Yorkshire,' *Proc R Soc Med*, 28 (1935), 603–16.

10 Underwood, E. Ashworth, 'The cholera epidemic in Exeter, 1832', *Brit Med J*, 1933 (i), 619–20; Mottram, R. H., in *Early Victorian England 1830–1865* (1934), ed, Young, G. M., in his chapter on 'Town life and London' (vol 1, 155–66), quotes freely from Shapter, T., *History of the Cholera in Exeter in 1832* (1849).

11 A facsimile of the original of this map is in Jusatz, H. J., 'Die geographisch-medizinische Erforschung von Epidemien,' *Peterm Geogr Mitt*, 86 (1940), 201–4.

12   In connection with Petermann's association with H. Berghaus and
A. K. Johnston it is worth noticing that the first maps of the
geographical distribution of diseases to appear in any German
atlas were in Band 2, Abt 7, no 2 (1847) of Berghaus, Heinrich,
Physikalischer Atlas (1837–48). The first similar map in England is
in Johnston, A. K., *The Physical Atlas of Natural Phenomena* (2nd
ed, 1856). The latter contains a world map of the geographical
distribution of health and disease (plate 35) with text by A. K.
Johnston; a red line traces the march of cholera from east to west,
with dates of its occurrence. This map does not appear in the first
edition (1848) of the atlas.

13   *Map of the British Isles elucidating the Distribution of the Population
based on the Census of 1841* (London, 1849). This map, compiled
and drawn by Augustus Petermann on a scale of 1: 1,600,000 was
exhibited to the statistical section of the British Association for
the Advancement of Science at Swansea in 1848; and Petermann
read a paper on the subject to the section. See *Geogr J*, 121 (1955),
448–50.

14   The official Report of the General Board of Health on the cholera
epidemic of 1849 includes a rough 'Cholera Map of the Metropolis, 1849', which indicates by very crude shading the districts of
London most affected; as well as a similar map of Glasgow. See
British Parliamentary Papers, 21 (1850). Dots showing deaths from
cholera in London in 1866 were drawn superimposed on 'maps
of the geological formations' in the *Report of Medical Officer for
the Privy Council for 1866* (London, 1867).

15   Acland, H. W., *Memoir on the Cholera at Oxford* (1856), 49–50.

16   Atlay, J. B., *Sir Henry Wentworth Acland* (1903), 197. Acland was
a lifelong friend of Ruskin's. His former home in Broad Street,
Oxford, became the university's School of Geography in 1910.

17   Haviland, A., *Scarborough as a Health Resort: its Physical Geography,
Geology, Climate and Vital Statistics with a Health-guide Map* (1883).

CHAPTER FIVE

Richard Ford (1796–1858)

AFTER THE END of the Napoleonic wars an annually increasing number of British tourists visited the Continent, but very few adequate English guidebooks were then available to assist these travellers.[1] It is true that an admirable guidebook to Switzerland by M. J. B. Ebel, first published in German in 1793, was brought out in an English version by Daniel Wall in 1818, and that useful guides to Belgium by Edmund Boyce, and to Italy by Mrs Mariana Starke appeared in 1815 and 1820.[2] In addition there were a number of inferior guidebooks for France and Germany and some eighteenth-century works with general information and advice for those about to make the Grand Tour. Mrs Starke's book is very discursive and contains a 'medley of classical lore, borrowed from Lemprière's Dictionary, interwoven with details regulating the charges in washing-bills at Sorrento and Naples, and an elaborate theory on the origin of *Devonshire Cream*, in which she proves that it was brought by Phoenician colonists from Asia Minor into the west of England'.[3] This book was useful because it included much pratical information gathered on the spot, not only about Italy, its main concern, but also about France, Switzerland, Germany, Portugal, Spain, the Netherlands, Denmark, Sweden and Russia.

The lack of accurate guidebooks was first made good on a large scale by John Murray (1808–92). In 1829, when he was still a young man, this publisher's son travelled in Holland and Germany. He set out for the Continent 'unprovided with any guide, excepting a few manuscript notes about towns and inns, etc, in Holland, furnished me by my good friend, Dr Somerville, husband of the learned Mrs Somerville'.[4] It appears that these notes were of the greatest value to him and that he felt the lack of them when he arrived in Hamburg. He remarked that the need for such friendly aid 'impressed on my mind the value of practical information gathered on the spot, and I set to work to collect for myself all the facts, information, statistics, etc, which an English tourist would be likely to require or find useful. I travelled thus, notebook in hand, and whether in the street, the *Eilwagen*, or the Picture Gallery, I noted down every fact as it occurred. These notebooks (of which I possess many dozens) were emptied out on my return home, arranged in Routes, along with such information as I could gather on history, architecture, geology and other subjects suited to a traveller's needs'.[5] On his return home the young man submitted his notebooks to his father, who had no previous knowledge of the scheme, but thought the work worth publication and 'gave it the name of "Handbook", a title applied by him for the first time to an English book'[6]. Previously the word 'handbook' had been applied to manuals of ecclesiastical offices, and it was now given to small books containing concise information for travellers. The firm of Murray did not publish a handbook immediately after John Murray's return from Germany. Murray tells us how his work was checked. 'Having drawn up my routes', he says, 'and having had them roughly set in type, I proceeded to test them by lending them to friends about to travel, in order that they might be verified or criticized on the spot. I did not begin to publish until after several successive journeys and temporary residences in Continental cities, and after I had not only traversed beaten routes, but explored various districts into which my countrymen had not yet penetrated'.

The first of Murray's handbooks was not published until 1836 and was entitled *Hand-book for Travellers in Holland, Belgium, Prussia, and Northern Germany, and on the Rhine from Holland to Switzerland*. This work was followed at short intervals by others on Southern Germany (1837), Switzerland, Savoy, and Piedmont (1838), and France (1843),[7] all three also written by Murray himself. The volume on France was based on four or five journeys made by Murray between 1830 and 1841. As the series proceeded

Plate 7   Richard Ford, from a sketch by J. F. Lewis in 1832

Plate 8 (a)   Ciudad Rodrigo seen from Alberga. Watercolour by Richard Ford

Plate 8 (b)   Talavera from the battlefield of July 1809. Watercolour by Richard Ford

it became clear that Murray could not carry out the whole scheme single-handed, and the handbook on Denmark, Norway, Sweden, and Russia (1839) was not written by Murray. In 1850, Anthony Trollope, not yet famous as a novelist, offered to write a handbook for Ireland in the series. At Murray's request he drafted some chapters, but his work was returned, unopened, nine months later.[8] Among the distinguished collaborators selected by Murray were Sir Gardner Wilkinson for Egypt, Sir Lambert Playfair for Algeria and for the Mediterranean, and Richard Ford for Spain. All three were at one time Fellows of the Royal Geographical Society. John Murray himself was one of the original Fellows and gave valuable service on the Council. At his death, in 1892, he was the last of the original Fellows of the Royal Geographical Society, established in 1830, still remaining on the list.

There were no English guidebooks to Spain, although a *Guide du Voyageur en Espagne*, by Bory de Saint Vincent was published in 1823. Spain was not included in the Grand Tour made by young aristocrats in the eighteenth century. Travel in Spain during that century, and even after 1815, was difficult and sometimes dangerous, and involved the harships usually associated with exploration. The roads were bad and often impassable except on horseback, while the inns were primitive. Many foreign merchants frequented the Spanish ports but only the most adventurous travellers penetrated into the interior of the Peninsula in the pre-railway age.

In 1839 Richard Ford was invited to dine with John Murray. During the evening the host asked his guest to suggest the name of a suitable author to write a handbook of Spain for his series. Ford replied, almost in jest that he would do it himself. In 1840 Ford was formally approached by Murray and agreed 'to do a handbook for Spain' and made a rash guess that he would complete 'the little book' in six months.[9] In order to understand why John Murray asked Richard Ford to write on Spain, something must be said about his early life and his interest in the Peninsula. Ford, who was born in 1796, was the eldest son of Sir Richard Ford, a friend and supporter of William Pitt, and at one time chief police magistrate at Bow Street, and creator of the mounted police force. Sir Richard, disguised as an old gentleman, once captured a highwayman on Hampstead Heath, an exploit which gained him great favour with George III.[10] The young Richard Ford was educated at Winchester, where he was contemporary with Arnold, and at Trinity College, Oxford, where he took his degree in 1817. He was called to the bar at Lincoln's Inn, but never practised. He married, in 1824, Harriet Capel, a daughter of the Earl of Essex, and, in 1830, he took her and his family to the south of Spain in order that her health might benefit by a residence in a warmer climate. In November of 1830 the Ford family arrived at Seville, where they lived for six months. In the following April Ford visited Madrid, the journey from Seville lasting four and a half days, and returned by way of Talavera, Mérida, and Badajoz. In June 1831, the family removed to Granada, where they resided in the palace of the Alhambra. From Granada Ford and his wife made the ascent of the Picacho de la Veleta (11,158ft), one of the highest peaks in the Sierra Nevada. In September Ford made an expedition into eastern Spain. Mrs Ford rode on a donkey while her husband was on horseback and their servant drove a covered cart. They proceeded by way of Elche to Alicante, Valencia, Murviedro, Barcelona, Monserrat, Manresa, Cardona to Zaragoza, and eventually to Madrid, returning to the Alhambra in November. In February 1832, Ford went on a riding expedition to Tarifa, Algeciras, Jerez and Ronda. Later in the same year he spent two months on a journey along the frontiers of Spain and Portugal, visiting some of the battlefields of the Peninsular War, and also to the north of Spain. He went to Mérida, Alcántara, Plasencia, Salamanca, and Ciudad Rodrigo, and later to Benavente, Santiago, Oviedo, León, and so to Bilbao and the Basque country, returning to Madrid by way of Burgos and Valladolid. The artist J. F. Lewis spent about three months with Ford at Seville during the winter of 1832–3.[11] The portrait of Ford reproduced in this chapter (Plate 7) was made by Lewis at this time. In 1833 Ford made a trip across the Strait of Gibraltar to Tetuan, was in Madrid in September, and finally returned to England in December. Two years later George Borrow made his first journey to the Peninsula.

The remarkable fact is that Ford spent only three years in Spain and in that comparatively short time he acquired an amazing knowledge of every aspect of the country's life and history. He travelled mainly on horseback and rode more than 2,000 miles on his favourite horse, Jaca Cordovese. In this way he penetrated into the most remote and inaccessible parts of Spain, and, facing every inconvenience with good temper, he stored in his notebooks the results of his adventures, combined with information for artists, historians, sportsmen, and students of dialects and customs. He consorted equally with grandees, bandits, and smugglers, and was thus able to understand and to appreciate the strange land of Spain perhaps better than any foreign traveller before or since.

On his return from Spain, Ford settled at Heavitree near Exeter and occupied himself in literary work of various kinds. In 1837 Murray published an anonymous pamphlet, *An Historical Enquiry into the Unchangeable Character of a War in Spain*. This is known to have been the work of Richard Ford and is a trenchant and full-blooded attack on another political pamphlet, *The Policy of England towards Spain*, published earlier in the same year in support of Lord Palmerston. Although Ford's pamphlet is mainly concerned with the 'disastrous condition of our foreign policy, as regards the Peninsula', it also reveals the immense knowledge which he had accumulated of all things Spanish. Ford's articles on Spanish subjects in the *Quarterly Review*, notably one on 'Spanish bull-feasts and bull-fights',[12] published in 1838, gave him a growing reputation and led eventually to Murray's invitation to write a handbook on Spain. The writing of the handbook occupied Ford for five years. He worked patiently in his garden-house at Heavitree, generally clad in his black jacket of Spanish sheepskin. There were many hitches, and he became at times heartily weary of the mechanical side of the work: all later writers of handbooks can sympathise with him. The handbook went to press in 1844 but the edition was cancelled and only twenty copies now exist: it ends abruptly at p 768. Ford took this drastic step on the advice of his friends who believed that some of his remarks were too caustic. He re-wrote a considerable part of the book before the first edition was finally published in two volumes in the summer of 1845. The original edition had been reprinted '*minus political, military and religious discussions*', as Ford explains in a letter.[13]

During the composition of his handbook Ford made the acquaintance of George Borrow. The latter had arrived in Portugal in November 1835, two years after Ford had left Spain, and spent the best part of the next five years in the Peninsula. Ford did not meet Borrow until 1841 after he had advised Murray to publish Borrow's *The Zincali; or an Account of the Gypsies of Spain*. Ford was also one of the first to read Borrow's famous work *The Bible in Spain*, which was published in December 1842, about two years before the handbook. Ford told Murray that *The Bible in Spain* would sell well and he later reviewed it in the *Edinburgh Review*,[14] receiving £44 in payment which he invested in Château Margaux.[15] Borrow in a letter to Murray says that he had seen 'the article in the *Edinburgh* about the Bible—exceedingly brilliant and clever, but rather too epigrammatic, quotations scanty and not correct. Ford is certainly a most astonishing fellow; he quite flabbergasts me—handbooks, reviews, and I hear that he has just written a *Life of Velasquez*, for the *Penny Cyclopaedia*'.[16] For some years the two literary travellers, Borrow and Ford, corresponded on terms of friendship and visited each other. On one occasion Borrow consulted Ford about the possibilities of travel in Africa.[17]

The full title of the handbook is *A Hand-book for Travellers in Spain and Readers at Home. Describing the country and cities, the natives and their manners; the antiquities, religion, legends, fine arts, literature, sports and gastronomy: with notices on Spanish History*. The first edition of the book is made up of fourteen sections, the first of which, called Preliminary Remarks, covers 142 closely printed pages. This section is divided into twenty-four parts and contains much that is still of value to the student of the geography of Spain. Ford makes some shrewd remarks on climate and attempts a division into climatic regions, and he gives detailed accounts of agricultural life. For example, the manufacture of olive oil and of sherry wine are carefully described. Throughout the book Ford emphasises the importance of regional differences in Spain and on the first page of his Preliminary Remarks he writes: 'The general comprehensive term "Spain", which is convenient for geographers and politicians, is calculated to mislead the traveller. Nothing can be more vague or inaccurate than to predicate any single thing of Spain or Spaniards which will be equally applicable to all its heterogeneous component parts. The north-western provinces are more rainy than Devonshire, while the centre plains are more calcined than those of Barbary: while the rude agricultural Gallician, the industrious manufacturing artisan of Barcelona, the gay and voluptuous Andalucian, are as essentially different from each other as so many distinct characters at the same masquerade'. He explains that 'it will therefore be more convenient to the traveller to take each province by itself and treat it in detail'.[18] The remaining thirteen sections of the book are given over to Andalusia, Ronda and Granada, the Kingdom of Murcia, Valencia, Catalonia, Extremadura, León, the Kingdom of Gallicia, the Asturias, the Castiles, Old and New, the Basque Provinces, the Kingdom of Arragon and the Kingdom of Navarre. Each of these sections is prefaced 'with a few preliminary remarks' giving 'those social and natural characteristics which particularly belong to each division, and distinguish it from its neighbours'. These 'few preliminary remarks' cover sixty-two pages in the section on Andalusia alone. In a letter to Addington, Ford wrote: 'I am sick of Handbook. I meditate bringing out the first volume, the *preliminary* and most difficult, early next spring. It is nearly completed. It is a series of essays, and has plagued me to death. The next volume will be more mechanical and matter-of-fact—what Murray wanted; and I am an ass for my pains. I have been throwing pearly articles into the trough of a road-book'.[19] The introductory information for each division of Spain is followed by a series of itineraries. There are nine routes for Andalusia, one of which is a 'mining tour' to Río Tinto and Almadén. But even the intineraries of the handbook are not written in the dull mechanical style of the normal road-book. The routes are packed with incidents and with references to Spanish art, literature, and history. Ford was a devoted student of Don Quixote and carried a copy with him on all his travels in Spain. He entreated the traveller 'to arm himself beforehand with a Don Quixote'.[20] He also makes many references in his routes to the Peninsular War and to the Duke of Wellington, of whom he was a great admirer and whose despatches he quotes freely.

It will be noticed that 'gastronomy' appears in the full title of the handbook and cookery seems to have been one of Ford's liveliest interests; his salads as well as the *ollas* and *guisados* of his table were famous. He devotes a chapter of the later *Gatherings from Spain* to Spanish cookery while the itineraries of the handbook are full of references to local dishes. He himself was responsible for introducing Montanches hams as well as Amontillado sherry into England. After describing the delicious Montanches hams he goes on to give a geographical explanation of the consumption of so much dried food in Spain. 'The fat of Montanches hams,' he writes, 'when they are boiled, looks like melted topazes, and the flavour defies language, although we have dined on one this very day, in order to secure accuracy and inspiration. . . . The nomad habits of Spaniards require a provision which is portable and lasting; hence the large consumption of dried and salted food, *bacalao*, *cecina*, etc. Their backward agriculture, which has neither artificial grasses nor turnips, deprives them of fresh meats and vegetables during many months; hence rice and *garbanzos* supply green herbs and appropriately accompany salted fish and bacon'.[21]

There are 140 itineraries in the book, which has 1,012 pages of text and further fifty pages of index. The two volumes of the first edition are very unattractive in appearance; the itineraries are printed in double columns. Yet, in spite of its small type, its high price (30*s*), and its unpretentious title, the book was an immediate success, no less than 1,389 copies being sold in three months. The book contains few statistics. Ford described Spain as 'no paradise for calculators', and even the fashionable quantifiers of today would find Spanish statistics a tough problem.

In 1846 Ford's *Gatherings from Spain* was published by Murray. This book, which had been prepared at great speed, consists partly of the introductory essays to the handbook and partly of new material. The book was praised by the critics and was a successful venture. Ford remarks that 'I have sacked £210 by two months' work and not damaged my literary reputation'.[22] This book is a very readable and lively account of Spain, and, as it was reprinted in Everyman's Library in 1906, is probably better known than the handbook on which it is largely based.[23] John G. Lockhart wrote to congratulate Ford on his achievement. 'You may', he says, 'live fifty years without turning out any more delightful thing than the *Gatherings*. Tho' I had read the *Handbook* pretty well, I found the full zest of novelty in these Essays, and such, I think, is the nearly universal feeling.'[24] A tribute to Ford's *Gatherings from Spain* was made by an American author, T. J. Hamilton in 1943. In a bibliographical note he says, 'For the reader who has time for only one book on Spain, however, the work to be recommended above all others is Richard Ford's *Gatherings from Spain*. Few nations have been so fortunate in their interpreters. Ford travelled for years in a land laid waste by war, and his account of the misery which bad government brought to a superb people is as valid to-day as it was when it was published a century ago'.[25]

Ford supervised the publication of a second edition of his handbook which appeared in 1847, this time in one volume. A third edition in two parts, published in 1855, was the last edition prepared by Ford himself, but numerous editions appeared after Ford's death in 1858. The fourth was published in 1869 and the ninth in 1898.[26] Most of Ford's other literary work is to be found in periodicals, notably in the *Quarterly Review* and the *Edinburgh Review*. Many of his articles in these journals are devoted to the literature, art, and architecture of Spain, subjects on which Ford had become an authority. He was elected a Fellow of the Society of Antiquaries in 1850. He had written a biography of Velasquez in the *Penny Cyclopaedia* in 1843, and he has been acclaimed as 'the first to unveil the greatness of Velasquez to northern eyes'. In an article entitled 'Apsley House', published in the *Quarterly Review* in 1853, he paid a tribute to the Duke of Wellington, with whom he appears to have had friendly personal relations.[27] When living in Spain at Granada, Ford was only ten miles from an estate of 4,000 acres which had been conferred by the Cortes on the Duke of Wellington after Salamanca. Ford had arranged that the Duke should receive his income in England, while he drew the same sum from the Duke's Spanish estate, thus overcoming difficulties of exchange. In 1852 Ford produced *A Guide to the Grand National and Historical Diorama of the Campaigns of Wellington*, a book which is mainly illustrated by William Telbin but includes some pictures for which the sketches had been made by Ford himself (Fig 14). Ford was an enthusiastic amateur artist and made numerous sketches in Spain, as well as a number of watercolours. (Plates 8 (*a*) and 8 (*b*).) Ford had previously supplied his friend, David Roberts, with sketches of Salamanca and other Spanish towns for a book called *The Tourist in Spain* (1838). Two other illustrated books, the texts of which were written by Ford, are *Tauromachia or the bull-fights of Spain* (1852)[28] and *Apsley House and Walmer Castle* (1853). Ford's last literary work was a review of *Tom Brown's Schooldays* in the *Quarterly Review* for October 1857: Ford's niece had married Thomas Hughes. In December 1856 Ford was appointed one of the Royal Commissioners to decide upon a site for the National Gallery but resigned on account of his health before the Commission had actually met. He died on 31 August 1858, and was buried in the churchyard at Heavitree, near Exeter; on his tomb he is fitly described as RERUM HISPANIAE INDAGATOR ACERRIMUS.

Ford's fame rests largely on his handbook. From its publication in 1845 until the present day all who have known Spain intimately have given it their praise. 'Surely never was there', wrote Prescott, 'since Humboldt's book on Mexico, such an amount of information, historical, critical, topographical, brought together in one view and that in the unpretending form of a *Manuel du Voyageur*'. Perhaps the best tribute paid to the handbook is that written by the third Earl of Carnarvon in the third edition of his *Portugal and Gallicia*. Speaking of Ford he says, 'he is imbued with that spirit of the land, which no man can successfully evoke unless his mind be fitted by something of a kindred nature to receive the spell. He has transferred to his own glowing but accurate pages that vivid appreciation which he so singularly possesses of all that is characteristic in Spain. Spain lives in his book, clad in her own peculiar and inimitable colouring'.[29]

Under the humble name of a handbook Ford had made 'a permanent contribution to literature';[30] he had produced one of the best books of travel written in the English language and the Spain of the pre-railway era will live for ever in its pages. Moreover, Ford's graphic picture of Spain is still true in its essentials: Spain changes but little. But Ford's handbook is not only a guidebook for Spain; it is a guide to the art of travel in any country, for it shows how a rich variety of interests combined with a zest for enjoyment can make the perfect traveller.[31]

Fig 14   Badajoz. Drawing by Richard Ford from his Guide to the Diorama of the Campaigns of Wellington *(1852)*

NOTES

1   This chapter was first printed in *Geogr J,* 106 (1945), 144–51, under the title *Richard Ford and his Hand-book for Travellers in Spain* to mark the centenary of the publication of the Hand-book. The chapter was reprinted in *Book Handbook,* no 6 (1948), 351–68, with some alterations and additional illustrations kindly lent by Brinsley Ford, great grandson of Richard Ford.

2   Ebel, M. J. G., and Wall, Daniel, *The Traveller's Guide through Switzerland* (1818); Boyce, Edmund, *The Belgian Traveller, or a Complete Guide Through the United Netherlands* (1815); Starke, Mariana, *Travels on the Continent, written for the Use and Particular Information of Travellers* (1820) known in its sixth ed (1828) as *Information and Direction for Travellers on the Continent.*

3   Murray, John, 'The origin and history of Murray's handbooks for travellers,' *Murray's Magazine*, 6 (1889), 624.

4   Murray, John, ibid, 624. See also Smiles, S., *Memoirs and Correspondence of the late John Murray* (2 vols, 1891), especially chap 35 on 'Murray's Handbooks'. For an account of Mrs Somerville see Baker, J. N. L., 'Mary Somerville and Geography in England', *Geogr J*, 111 (1948), 207–22.

5   Murray, John, ibid, 624.

6   ibid, 625.

7   The third edition (1848) of Murray's *Hand-book* on France was used extensively by John and Effie Ruskin on their journey to northern France in 1848. See Links, J. G., *The Ruskins in Normandy. A Tour in 1848 with Murray's Hand-book* (1968). This is a fascinating account of travel at a time when railways were spreading at the expense of the diligences.

8   Trollope, A., *An Autobiography* (1883); in World's Classics ed, (1923), 79–80.

9   Prothero, R. E., *The Letters of Richard Ford 1797–1858* (1905), 172–3 in a letter dated 4 August 1839.

10 'Richard Ford. In Memoriam', *Fraser's Magazine*, 58 (1858), 422–4

11 Ford, B., 'J. F. Lewis and Richard Ford in Seville, 1832–33', *Burlington Magazine*, 80 (1942), 124–9.

12 *Quarterly Review*, 62 (1838), 385–424.

13 Prothero, R. E., *The Letters of Richard Ford*, op cit, 196 in a letter dated 19 February 1845.

14 *Edinburgh Review*, 77 (1843), 105–38.

15 Prothero, R. E., op cit, 186 in a letter dated 27 February 1843.

16 Smiles, S., *Memoirs of John Murray*, op cit, 2, 493, in a letter dated 25 February 1843.

17 Shorter, Clement K., *The Life of George Borrow* (1919), 191. This book gives an interesting account of Ford's correspondence and relations with Borrow.

18 Ford, R., *Hand-book for Travellers in Spain*, 1 (1845), 1.

19 Prothero, R. E., op cit, 178–9 in a letter dated 18 November 1841. The author of this book wrote this chapter on Ford during the war, when he was compiling and editing four volumes for the Admiralty in the *Geographical Handbook Series* on *Spain and Portugal*

(1941–5). Like Ford he was 'sick of Handbook' when he wrote this essay!

20 Ford, R., *Hand-book*, op cit, 1 (1845), 307.

21 ibid, 1 (1845), 545.

22 Prothero, R. E., op cit, 204 in a letter dated December 1846.

23 A further reprint of *Gatherings from Spain* was produced in Every-man's Library in 1970 with an introduction by Brinsley Ford.

24 Prothero, R. E., op cit, 204–5 in a letter dated 5 January 1847.

25 Hamilton, Thomas J., *Appeasement's Child. The Franco Regime in Spain* (1943), 238.

26 A new reprint of the *Hand-book* was produced in three vols (Centaur Press ed, 1966) with a valuable introduction by Ian Robertson, and a foreword by Sir John Balfour.

27 *Quarterly Review*, 92 (1853), 446–86.

28 Stirling-Maxwell, W., *Miscell. Essays and Addresses includes* 'Spanish bull-fights', vol 6 (1891), 235–55, which is a review of Price, L., and Ford, R., *Tauromachia*. Ford's account of a bull-fight is quoted at some length.

29 Herbert, H. J. G., third Earl of Carnarvon, *Portugal and Gallicia* (3rd ed, 1848), viii.

30 D.W.F.[reshfield] in an obituary of John Murray, *Proc RGS*, 14 (1892), 334.

31 The following articles are useful for students of Ford: Ford, Brinsley, 'Richard Ford's articles and reviews', *Book Handbook* (1948, no 7), 369–80 with a complete bibliograpny of Ford's writings; Ford, Brinsley, 'Spain in the 1830's', *Geog Mag*, 21 (1958), 159–70 illustrated by many of R. Ford's sketches; Hoskins, W. G., 'The finest travel-book in English', *The Listener*, 60 (no 1536, 4 September 1958), 237–9.

CHAPTER SIX

British Regional Novelists and Geography*

IN MANY COUNTRIES the prophets of the idea of the region have been poets and novelists: this is especially the case in England. A very discerning Indian observer recently remarked: 'English literature is the best guide for foreigners to the English scene because it is more closely the product of its geographical environment, more ecological, than any other literature I have read. I think English literature has gone farthest in fusing Nature and the spirit of man'.[1]

One aspect of the general nineteenth-century revival of interest in locality lay in the output of regional novels in European literature. The novelists, and not the historians, the archaeologists and the geographers, were the first to make a detailed record of the features which colour a locality. English literature is extremely rich in regional novels, partly because England is a very diverse country from a geographical point of view. American visitors to England are often impressed by the great diversity of scenery within very small areas. It is not surprising that geology as a science really began in England. There are in England numerous well marked regions, each of which, while forming part of the

* This chapter originally appeared on pp 47–54 of the following book, and is reproduced with the kind consent of its editors and publishers: Abhandlungen des 1. Geographischen Instituts der Freien Universität Berlin (Neue Folge der Abhandlungen des Geographischen Instituts der Freien Universität Berlin). *Band 13, Aktuelle Probleme Geographischer Forschung. Festschrift für Joachim Heinrich Schultze aus Anlass seines 65. Geburtstages.* Herausgegeben von Klaus-Achim Boesler und Arthur Kühn. Mit 43 Photos und 66 Figuren, davon 4 auf 2 Beilagen. (Verlag von Dietrich Reimer in Berlin, 1970).

national entity, possesses a distinct character of its own. The material for English regional literature therefore exists in abundance.

Professor Lucien Leclaire, Professor of English Language and Literature in the University of Caen has written two remarkable books on British regional novels.[2] Professor Leclaire is not a geographer but he illustrated his books with some interesting maps. On one of these maps he shows the distribution of regional novelists of England and Wales—that is of those who wrote between 1800 and 1950 (Fig 15). The names of over 150 novelists were placed, as far as was possible, on the part of the map which was the setting of their regional novels.[3] Professor Leclaire also compiled two maps on which are underlined the names of regional novelists in whose books an industrial rather than a rural background is dominant. Professor Leclaire argued that between 1870 and 1910 by far the main interest of the novelists lay in rural regions but that from 1910 the pendulum swung in the other direction; and that between 1940 and 1950 novels with an industrial background were as numerous as those with a rural scene.

From 1800 until about 1830 the forerunners of the English regional novel have a national rather than regional flavour. Mr J. H. Paterson, in a recent article on Sir Walter Scott (1771–1832) argues that Scott's novels were more national than regional[4] Paterson states that Scott was 'essentially a historical novelist, moving from period to period and so preserving no chronological unity but, even when dealing with a particular period or theme, he ranged over the length and breadth of the country, with all its diversity of landscapes', and added that 'Scott's writing never remained attached to one place . . . for long enough to do for it what Hardy did for Dorset'. Nevertheless Paterson rightly considers that Scott deserves a place as a pioneer in the development of the regional novel. He shows that Scott paid very great attention to topographic detail in the background of his stories and he argues that 'for thousands of people both during his lifetime and after his death, it was his writing which provided their introduction to the Scottish landscape'.

From about the time of Sir Walter Scott's death in 1832 until,

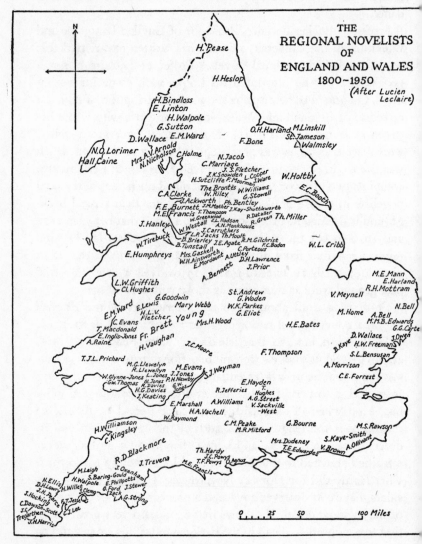

THE
REGIONAL NOVELISTS
OF
ENGLAND AND WALES
1800~1950
(After Lucien Leclaire)

H. Pease

H. Heslop

H. Bindloss
E. Linton
H. Walpole
G. Sutton
D. Wallace   E.M. Ward
N.O. Lorimer   Mrs. A.V. Arnold
Hall Caine   N. Nicholson   C. Holme
O.H. Harland   M. Linskill
St. Jameson
F. Bone   D. Walmsley
N. Jacob
C. Marriage
J.S. Fletcher
J.K. Snowden   L. Cooper
H. Sutcliffe   F. Moorman   J. Ware
The Brontës   W. Williams
W. Riley   G. Stowell
C.A. Clarke   Ph. Bentley
J. Ackworth   J.H. Mather   J. Kay-Shuttleworth
F.E. Burnett   T. Thompson   R. Dataller
M.E. Francis   W. Greenwood   J.L. Hodson   R. Green   Th. Miller
J. Hanley   W. Westall   A.N. Monkhouse
J. Carruthers
W. Tirebuck   J.P. Jacks Th. Moult   R.M. Gilchrist
B. Brierley   J.E. Agate   F.C. Boden
E. Humphreys   B. Tunstall   C. Porteous
Mrs. Gaskell   A. Utley   W.L. Cribb
W.H. Ainsworth   D.H. Lawrence
E. Mordaunt   J. Prior
A. Bennett
M.E. Mann
E. Harland
L.W. Griffith   R.H. Mottram
Cl. Hughes   G. Goodwin   St. Andrew   V. Meynell
E. Lewis   Mary Webb   G. Woden   M. Home   N. Bell
E.M. Ward   H.L.V.   W.K. Parkes   A. Bell
C. Evans   Fletcher   G. Eliot   M.M.B. Edwards
T. Macdonald   Fr. Brett Young   Mrs. H. Wood   H.E. Bates   G.G. Carte
E. Inglis-Jones   D. Wallace   J. Owen
A. Raine   H. Vaughan   B. Kaye   H.W. Freeman
J.C. Moore   F. Thompson   S.L. Bensusan
T.J.L. Prichard   A. Morrison
M.G. Llewelyn   M. Evans   C.E. Forrest
R. Llewellyn   J. Jones   S.J. Weyman
W. Glynne-Jones   L. Jones   P.H. Newby   E. Hayden
Gw. Thomas   Gl. Jones   G.W.S.   R. Jefferies   Hughes
H.G. Davies   A. Davies   Jones   A.G. Street
J. Keating   E. Marshall   A. Williams   V. Sackville
H.A. Vachell   -West
W. Raymond
C.M. Peake   G. Bourne   M.S. Rawson
H. Williamson   M.R. Mitford   S. Kaye-Smith
C. Kingsley   Mrs. Dudeney   V. Brown   O. Ollivant
R.D. Blackmore   T.E. Edwardes
Th. Hardy
J. Trevena   J. Oxenham   J.C. Powys   O. Onus
M. Leigh   M.E. Francis   T.F. Powys
N. Ellis   S. Baring-Gould
D.H. Lowry   H. Walpole   E. Phillpotts   J. Stewer
J. Hocking   J.H. Pearce   Clemo   G. Ford   Zack   L.A.G. Strong
C. Dawson   Scott
J.C. Tregarthen   C.L. Lee
J.H. Harris

0   25   50   100 Miles

Fig 15

say, 1870, the English novel became more narrowly local, while during the eighty years after 1870 the regional novel was developed in a number of different forms. Leclaire's distribution map shows that authors were writing about many parts of England and Wales but more of them were dealing with western regions, Cumbria, the Pennines, Wales, Cornwall and Devon than with the midland and eastern parts of the country. In the opinion of Phyllis Bentley, the 'four great writers of the past who created, developed, and possibly perfected the English regional novel' are Charlotte Brontë (1816–55), George Eliot (1819–80), Thomas Hardy (1840–1928), and Arnold Bennett (1867–1931).[5] Charlotte Brontë's classic regional novel *Shirley* was published in 1849. The author lived in the moorland village of Haworth in West Yorkshire and the natural surroundings had a deep influence on her character and writing. She saw the beauty of nature but she also observed its fierce power in the winds and streams of this wild region. When Mrs Gaskell met Charlotte Brontë at Windermere in the Lake District in August 1850 she wrote: 'I was struck by Miss Brontë's careful examination of the shape of the clouds and the signs of the heavens, in which she read, as from a book, what the coming weather would be, I told her that I saw she must have a view equal in extent at her own home. She said that I was right, but that the character of the prospect from Haworth was very different; that I had no idea what a companion the sky became to anyone living in solitude,—more than any inanimate object on earth,—more than the moors themselves'.[6] The setting of *Shirley* is in the woollen district of the West Riding of Yorkshire. Most of the characters in the book gain their livelihood from the textile industry which is so intimately connected with the physical geography of the region. The conversation and dialogue of the book are often in dialect, that is in regional speech. In Emily Brontë's (1818–48) *Wuthering Heights* (1847) the setting is regional and there are magnificent descriptions of the moors of the West Riding but the plot is not so tied to place as in *Shirley*. Although *Wuthering Heights* is not so regional as *Shirley* it is probably a greater novel. Mrs Gaskell (1810–65) wrote two novels about industrial Lancashire that can be classed as regional: *Mary Barton*

(1848) and *North and South* (1855). In both books dialect is freely used. *North and South* contrasts the peaceful life of a rural county in the south of England with the 'almost brutal strenuousness of the manufacturing north' as the author had seen it in 'Milton' that is Manchester.

Anthony Trollope (1815–82) was not a regional writer because his places were not real. He created a new county, Barsetshire, which looks very much like 'a county in the west of England . . . its green pastures, its waving wheat, its deep and shady and—let us add—dirty lanes, its paths and stiles, its tawny-coloured, well-built rural churches, its avenues of beeches, and frequent Tudor mansions, its constant county hunt, its social graces, and the general air of clanship which pervades it, has made it to its own inhabitants a favoured land of Goshen'.[7] In *The Warden* (1855), the first of the Barchester series, Trollope created the fictitious cathedral town of Barchester. He said that 'were we to name Wells or Salisbury, Exeter, Hereford or Gloucester, it might be presumed that something personal was intended'. It is possible to draw a map of Barsetshire because Trollope's fictitious topography was consistent.[8] Trollope himself made a map of the county 'the new shire which I had added to the English counties'. Trollope did not describe real places, but the convention of a fictitious topography was his idea and it was followed by George Eliot and Thomas Hardy. But these two authors described real places and districts to which they merely gave different names. Thus Hardy described Oxford in some detail but gave the city the name of Christminster, and its suburban district of Jericho he renamed Beersheba.

George Eliot's first four novels are regional. *Scenes of Clerical Life*, first published in *Blackwood's Magazine* in 1857, described the Warwickshire environment in which the author spent her early life. 'Milby' is Nuneaton and was easily recognised as such by readers at the time of publication. The action in *Adam Bede* (1859) takes place in 'Loamshire' (Staffordshire) and 'Stonyshire' (Derbyshire). In *The Mill on the Floss* (1860) the original of George Eliot's 'red-roofed town of St Oggs' is Gainsborough on the Trent, the town in which Sir Halford Mackinder was born on 15

February 1861, only a few months after the appearance of the novel. Mackinder often used to refer to George Eliot's connection with his birthplace, and to the tidal eagre, the 'crest-fronted wave' coming up the Trent ('Floss'). In *Silas Marner: The Weaver of Raveloe* (1861) the setting is in Warwickshire once more, 'Raveloe' being Bulkington, near Nuneaton. The region of George Eliot is largely agricultural, as is the Wessex of Thomas Hardy. The fourteen novels by Hardy, as well as some volumes of short stories, were published between 1871 and 1896. All these books have their setting in Wessex, the six counties in the south-west of England. Hardy states 'that the area of a single county did not afford a canvas large enough' for his purpose; nevertheless his 'South Wessex' (Dorset) possesses a well-marked individuality. Hardy portrays every aspect of this lovely landscape with complete fidelity. His detailed observation of the scene is matched by his knowledge of the history and traditions of the region. Miss Bentley believes that 'in Hardy's novels the local colour dyes the setting element more deeply, more richly, than in any other regional fiction'.[9] In a valuable paper, by means of skilfully chosen quotations from Hardy, Professor H. C. Darby has described the five divisions of Dorset as they appear in the novels: the chalk upland, the vale of Blackmoor, the heathland and heathland valleys, the Isle of Purbeck, and western Dorset.[10] Hardy devised new place-names for the whole of the large area of Wessex; in the later editions a map with these names is always included. The community described by Hardy was almost entirely agricultural and lived mainly in villages. Hardy's 'Weatherbury' (Puddleton) is a perfect example of the nucleated settlement, with its farms, inn, malthouse, church and great monastic barn. It must be remembered that the Wessex of Hardy's novels is a landscape of the nineteenth century and that it has been greatly changed in the years since Hardy wrote. The heart of the Hardy country is Dorset. F. W. Morgan was right when he said that 'of the best of the novels only one is set outside Dorset, and the Berkshire downland of *Jude the Obscure* is very similar to that of Dorset, while in the whole series the author's touch seems to become less sure whenever he moves beyond the region he knows so intimately'.[11] It is satisfactory to record that in August 1959, the so-called 'Hardy country', an area of about 400 square miles in Dorset was officially designated as an 'area of outstanding natural beauty' by the then Ministry of Housing and local Government, and came under the authority of the National Parks Commission. The area thus received the protection against unsuitable or harmful development that is provided by Act of Parliament.

Arnold Bennett's novels deal with an urban region, the industrial area of the Potteries which he made known as the Five Towns. His five, Turnhill, Bursley, Knype-on-Trent, Hanbridge and Longshaw are easily identified as Tunstall, Burslem, Stoke-upon-Trent, Hanley and Longton, Fenton never forgave Bennett for omitting it. Bennett began his series with *Anna of the Five Towns* (1902) in which the scene is laid mainly in Burslem. The Five Towns are, in Bennett's words, 'mean and forbidding of aspect—sombre, hard featured, uncouth', but just as Hardy had described ploughland and heath, so Bennett writes with equal affection, of pot-banks and furnaces. 'Bursley', said Bennett, 'lies toward the north end of an extensive valley, which must have been one of the fairest spots in Alfred's England, but which is now defaced by the activities of a quarter of a million people . . . nothing could be more prosaic than the huddled red-brown streets; nothing more seemingly remote from romance. Yet be it said that romance is even here. . . . Look down into the valley . . . embrace the whole smokegirt amphitheatre in a glance, and it may be that you will suddenly comprehend the secret and superb significance of the vast Doing which goes forward below. Because they seldom think, the townsmen take shame when indicted for having disfigured half a county in order to live. They have not understood that this disfigurement is merely an episode in the unending warfare of man and nature and calls for no contrition'. The single industry of the Potteries made this region into an industrial and social unit. In a passage in *Whom God Hath Joined* (1906) Bennett brings out this unity: 'All around the horizon . . . the yellow fires of furnaces grow brighter in the first oncoming of the dusk. The immense congeries of streets and squares, of little houses and great halls and manufactories, of church spires and proud smoking chimneys and chapel towers mingle together into one wondrous organism that stretches and rolls unevenly away for miles in the grimy mists of its own endless panting'. The nineteenth century industrial community is the theme of Bennett's work—the Five Towns with 'all their vast apparatus of mayors and aldermen and chains of office, their gas and their electricity, their swift transport, their daily paper, their religions, their fierce pleasures, their vices, their passionate sports and their secret ideals . . .'. Bennett's graphic account of the 'wondrous organism' of the conurbation is a powerful aid to its geographical understanding.

Another industrial area, the Nottinghamshire coalfield is the setting of much of the work of D. H. Lawrence (1885–1930); for example *The White Peacock* (1911) and *Sons and Lovers* (1913).[12] The former is drawn from the farming life Lawrence knew so intimately on the border of Nottinghamshire and Derbyshire. But Lawrence understood the life of the mining village equally well as he was brought up in one. Lawrence wrote an essay in 1929 on 'Nottingham and the Mining country'.[13] In this he tells that he was born in Eastwood, a large mining village not far from Nottingham and only a mile from the Erewash, a brook which divides the county of Nottingham from that of Derby. In the same essay Lawrence describes the beauty of the countryside in which he grew up. He differentiated between the red sandstone landscape of Sherwood Forest on the east and the limestone country of Derbyshire on the west; he regarded the mines as 'an accident in the landscape'. Lawrence has been regarded by many as a successor to Hardy because he wrote such eloquent and beautiful descriptions of the rural landscape. Similarly the work of a lesser known author, E. C. Booth (1872–1954), especially his novel *Cliff End* (1908), has been compared with that of George Eliot. Booth described the Holderness district of Yorkshire and the city of Hull with fidelity and considerable use of dialect. He gave a picture of East Yorkshire and its rural people as they toiled, spoke and laughed which is so accurate and vivid that it remains a record of its region as it was before the two wars. In his novels Doncaster appears as 'Daneborough' and Hull as 'Hunmouth'. Another novelist of the rural scene is Mary Webb (1881–1927) who writes of Shropshire, a region lying between 'the dimpled lands of England and the gaunt purpled steeps of Wales'. In *Precious Bane* (1924) Mary Webb displays her close knowledge of life on the Welsh border. The 1928 edition of this book had an introduction by Stanley Baldwin (1867–1947), then Prime Minister, who was a devoted admirer of Mary Webb's writing. Baldwin represented Bewdley in Parliament and himself knew Webb's Shropshire intimately. Phyllis Bentley (born 1894) writes novels which give a vivid picture of the West Riding and its grim beauty; her *Inheritance* (1932) is especially notable for its fidelity to place. Winifred Holtby (1898–1935) wrote *Anderby Wold* (1923) a story of a farm in the East Riding described with living accuracy, but her classic is *South Riding: an English Landscape* (1936), the setting of which is also laid in the East Riding of Yorkshire, in the valley of the Hull, between Beverley and Hull. This book was completed only one month before the author's lamented death at the early age of thirty-seven.

English regional novelists display many merits that geographers can recognise and envy. The novel has illuminated the English landscape as brilliantly as any other art. Again many novelists present life and work on a clearly marked piece of land with truth. Reality is faithfully shown; it is not lost in the dim twilight of modern geographical jargon. The regional novelists have painted many pictures of real earth with the fine attention to detail of a Dutch work of art. 'The regional novel', said Phyllis Bentley, 'has a particular strength in the depiction of character, for the two great factors in the formation of character are heredity and environment, and in the regional novel characters are shown in their native environment, and surrounded by their families, that is their ancestors and their descendants.'[14] If a geographer regards his subject as the study of the earth as the theatre of human life he will find many links with the regional novelists. The task of the regional geographer resembles that of the regional novelist, for the regional geographer strives to integrate the multitude of seemingly disconnected facts about nature and man in the region he is describing. 'Geography', said Mackinder, 'presents regions to be philosophically viewed in all their aspects interlocked'.[15] Like the Ancient Greeks the modern regional geographer tries to see things as a whole. He often fails to achieve this aim and he is bound to envy the greater success of the regional novelist. The intense power of environment in men's lives has been clearly perceived by many writers who are not geographers. Somerset Maugham in his *Razor's Edge* (1944) said: 'Men and women are not only themselves; they are also the region in which they were born'.

The regional novelists have been able to produce a synthesis, 'a living picture of the unity of place and people', which so often eludes geographical writing. The geographer often speaks of the 'personality' of a region and this is exactly what some novelists have brought out so strongly: they have successfully revealed the individuality of a particular landscape. All this lends support to the conclusion that regional description can never be an exact science, in spite of its scientific appearance; and that it will achieve greater success by the use of artistic and subjective methods. The view that 'a science of regions is a vain dream' has been held by Professor John Leighley. He argued that 'literary art, not systematic description, is the proper medium of regional synthesis'.[16] One of the greatest masters of geography has described the way in which mere area acquires geographic individuality. These are the words of Vidal de la Blache: 'A geographical individuality does not result simply from geological and climatic conditions. It is not something delivered complete from the hand of Nature.... It is man who reveals a country's individuality by moulding it to his own use. He establishes a connection between unrelated features, substituting for the random effects of local circumstances a systematic co-operation of forces. Only then does a country acquire a specific character differentiating it from others, till at length it becomes, as it were, a medal struck in the likeness of a people'.[17] It is because regional novelists in England and Wales, as well as in many other countries have made their regions shine like 'medals struck in the likeness of a people' that regional geographers should read and re-read their works. British regional

novelists have been true pioneers in the art of descriptive geographical writing.

NOTES

1 Chaudhuri, Nirod C., *A Passage to England* (1959), 15.
2 Leclaire, Lucien, *Le Roman Régionaliste dans les Iles Britanniques
   1800–1950* (Clermont-Ferrand, 1954); and *A General Analytical
   Bibliography of the Regional Novelists of the British Isles 1800–1950*
   (Paris, 1954). A revised and more complete edition of the bibliography appeared in 1969.
3 The map, reproduced here as Fig 15, was first re-drawn and
   printed on p 165 of Gilbert, E. W., 'The idea of the region',
   Geography, 45 (1960), 157–75.
4 Paterson, J. H., 'The novelist and his region: Scotland through the
   eyes of Sir Walter Scott', *Scott Geog Mag*, 81 (1965), 146–52.
5 Bentley, Phyllis, *The English Regional Novel* (1941), 13.
6 Gaskell, E. C., *The Life of Charlotte Brontë* (1857); and p 310 in
   Everyman's Library edition (1908).
7 Trollope, Anthony, *Doctor Thorne* (1858), chap 1.
8 Sadleir, Michael, *Trollope: a Commentary* (1945), 161–3 for three
   maps of Trollope's Barsetshire; see also Cox, J. Stevens, *Hardy's
   Wessex: Identification of fictitious place names in Hardy's works* (St.
   Peter Port, Guernsey, 1970).
9 Bentley, Phyllis, op cit, 25.
10 Darby, H. C., 'The regional geography of Thomas Hardy's
   Wessex', *Geogr Review*, 38 (1948), 426–43; see also Brinkley,
   Richard, *Thomas Hardy as a Regional Novelist* (St Peter Port,
   Guernsey, 1968), a monograph written with special reference to
   The Return of the Native (1878).
11 Morgan, F. W., 'Three aspects of regional consciousness',
   Sociological Review 31 (1939), 79
12 For a valuable account of Lawrence's work in 'describing the
   essence of a place' and in 'perceiving and recording a real differ-
   entiation', see Spolton, L., 'The spirit of place: D. H. Lawrence
   and the East Midlands', *East Midland Geographer*, 5 (1970), 88–96.
13 Reprinted in Lawrence, D. H., *Selected Essays* (Penguin Books,
   1954), 114–22.

14 Bentley, Phyllis, op cit, 45.
15 Mackinder, H. J., 'Progress of geography in the field and in the study during the reign of His Majesty King George the Fifth', *Geogr Journal*, 86 (1935), 12.
16 Leighley, John, 'Some comments on contemporary geographic method', *Annals Assoc Amer Geographers*, 27 (1937), 131.
17 Vidal de la Blache, P., *The Personality of France*, transl by Brentnall, H. C. (London, 1928), 14.

CHAPTER SEVEN

Victorian Methods of Teaching Geography

IN THE EARLY nineteenth century the academic study of geography in Britain had fallen to its lowest ebb, and that in
spite of the voyages and journeys of the great explorers.[1]
Geographical education consisted largely of learning by rote the
names of places and products. Our unfortunate ancestors had to
get by heart long inventories of capes and islands, of countries
and cities. The 'Scientific Tables' of *The Students' Classical Guide
to the Sciences* (1818) are typical of this kind of study. From these
tables children learnt that the principal 'animals' of England were
'horses, wild fowls, sheep and bustards' and that while the
'government' of England was 'Monarchial Aristocratical', that of
Denmark was 'Absolute Monarchy' and that of Turkey 'Despotic'.
(Fig 16.) A small book entitled *References to Geography used at
Townhead School near Rochdale* (1803) contains similar lists of
provinces and their capitals, of rivers and lakes, all to be got by
heart by school children, but it does also include statistics of
population and even some facts about the rate of wages in
Hindoostan. Sometimes the learning of the names of capes and
bays was made easier by fitting them into lines of doggerel verse.
The *Geographical Guide: a poetical nautical trip round the island of
Great Britain* (1805) describes the coastline of Britain in lines such
as these:

> 'Round England and Scotland prepare for a trip,
> And whilst British Tars are unmooring the ship,
> We'll over a map of the isles take a glance,

GEOGRAPHY.

	SWEDEN.			DENMARK.	RUSSIAN EMPIRE.		PRUSSIA.	
KINGDOMS. ...... with their	SWEDEN.	NORWAY.	LAPLAND.		RUSSIA in Europe.	POLAND.		ENGLAND.
Ancient Names.....	Scandinavia.			Cimbria.	Sarmatia.	Sarmatia.		Britannia.
Square Miles......:	81,000	112,000	70,000	12,896	630,000		24,000	57,680
Population.........	3 Millions.	800,000	60,000	3 Millions.	36 Millions.	13 Millions.	8 Millions.	11 Millions.
Government.	Absolute Monarchy.	Absolute Monarchy.	Absolute Monarchy.	Absolute Monarchy.	Absolute Monarchy.	Absolute Monarchy.	Absolute Monarchy.	Monarchial, Aristocratical.
Sovereign in 1817...	Bernadotte.	Bernadotte.	Bernadotte.	Frederic VI.	Alexander.	Alexander.	Frederick William IV.	George III.
Religion.	Lutheran.	Lutheran.	Lutheran.	Lutheran.	Greek Church	Rom.Catholic	Lutheran.	Protestant.
Capital. Latitude. Longitude. Miles from London. Inhabitants	Stockholm. 59 21 N. 18 4 E. 895. 70,000.	Bergen. 60 0 N. 5 25 E. 620. 20,000.	Tornea. 65 50 N. 24 17 E.	Copenhagen. 15 40 N. 12 40 E. 595. 100,000.	Petersburg. 59 50 N. 30 19 E. 1260. 170,000.	Warsaw. 52 14 N. 21 1 E. 870. 100,000.	Berlin. 52 32 N. 13 22 E. 590. 130,000.	London. 51 31 N. 6 W. 590. 1 Million.
Principal Towns....	Gottenburg, Abo, Upsal, Carlscrona.	Christiansand Bergen, Aggerhuus, Drontheim.	Kola.	Wiburgh, Sleswick, Ripen, Stralsund, Odensee.	Moscow, Archangel, Riga, Croustadt, Wyburg, Revel.	Cracow, Mittau, Wilna.	Konigsberg, Breslaw, Dantzic, Thorn, Posen.	York, Leeds, Liverpool, Sheffield, Manchester, Hull, Bristol, Exeter.
Universities.........	Upsal, Abo, Lund.			Copenhagen, Kiel.	Petersburg.	Cracow.	Konigsberg, Posen.	Oxford, Cambridge.
Principal Mountains	Swucku, Kinekulle.	The Dofrefield, Ardanger, The Koehlen.	Mass of Mountain.	Heckla, volcanic in Iceland.	The Uralian, The Valday, The Caucasus. The Golden Ridge	The Carpathian.		The Cheviot, The Skiddaw, Helvellyn, Malvern, Chiltern.
Principal Rivers...	Dahl, Ulea, Kymene, Tornea.	Glomen, Dramme.	Tornea, Tano, Alten.	Guden, Eyder.	Wolga, Don, Neiper, Neister, Dwina.	Vistula, Wesel, Neiper, Bog.	Vistula, Memel, Pregel, Warta.	Thames, Severn, Tyne, Humber, Ouse, Trent, Medway.
Principal Lakes. ..	Wenner, Wetter, Mœlar.	Rands-Sion, Ojeren, Mioss.	Hernasbostaer Tornea, Lulea.		Onega, Ladoga, Ilmen, Peypus.		Sperling Sea Maner Sea, Gneserich.	Windermere Coniston, Ulswater, Derweutwater
Principal Capes, ..		North Cape, The Naze.						Lizard'sPoint Laud's End, Spurn Head.
Islands	Gothland, Oeland, Aland, Rugen.	The Lofoden, Wardhuys.		Greenland, Iceland, Zealand, Funen.	Cronstadt, Oesel, Dago, Nova Zembla			Wight, Thanet, Sheppy, Guernsey.
Animals.	Elks, Seals, Squirrels, Varying Hares, CastorBeaver	Bears, Hares, Elks, Lynxes, Gluttons, Ermines.	Rein Deers, Lynxes, Ermines, Sables.	Horses and Horned Cattle.	Bears, Foxes, Wolves, Hyænas, Lynxes.	The Bohac, Gluttons, Buffaloes, Boars & Elks	Bisons, Beavers, Lynxes, Bears, Foxes.	Horses, Wild Fowls, Sheep, Bustards.
Produce and Trade.	Silver, Iron, Copper, Lead Pitch, Hemp, Tobacco, Alum.	Fir, Oak, Pine, Ash, Elm, Yew, Hides, Cods, Herrings.	Fir, Marble, Pearls, Jasper, &c. Skins of Ermines, Sables, Martins.	Horned Cattle, Horses, Tallow, Oil, Timber, Iron,Hides.	Tallow ,Oil, Hemp, Timber, Hides, Tar, Flax, Iron, Copper.	Corn, Salt, Flax, Hemp, Planks, Tar, Honey.	Glass, Paper, Brass, Iron, Gunpowder, Cloth, Amber, Honey, Hemp, Linseed.	Corn, Coals, WoollenCloth Hardware, Earthenware Carpets, Glass, Tin, Copper.
	SWEDEN.	NORWAY.	LAPLAND.	DENMARK.	RUSSIA.	POLAND.	PRUSSIA.	ENGLAND.

• For Celebrated Characters, see Biographical Table.

Fig 16   Scientific Tables for Geography from The Student's Classical Guide to the Sciences (*1818*)

Then start from Land's End, and sail round by Penzance.
Mount's Bay having cross'd, by the Lizard then steer
Your course to—North-East and by North quickly veer'
and continues in the same strain the whole way round Great
Britain with a mention of every cape and bay.

Another method of learning geography was known as the use
of the globes. (Fig 17.) As early as 1644 John Milton, a vigorous
believer in the study of geography, which he regarded as 'both
profitable and delightful', advocated the use of the globes in an
essay on education.[2] W. M. Thackeray in his *Vanity Fair*, pub-

E'en half the SCHOOL Authors be Read it will be seasonable for YOUTH to learn the Use of the GLOBES.

MILTON on Education.

Fig 17   The Description and Use of the Terrestrial Globe by Richard Turner from his View of the Earth: being a short but comprehensive System of Modern Geography *(1762), with a quotation from John Milton (1644)*

lished in 1847, gibed at Miss Pinkerton's Academy for young ladies because that institution advertised geography and the use of globes as a principal subject in its curriculum. Sir John Ross, in 1835, when describing an interview with an Eskimo woman called Tiriksiu,[3] said: 'I soon found too, that this personage, woman though she was, did not want a knowledge of geography, and that also of a different nature from what she might have acquired in an English boarding-school, through the question book and "the use of the globes". She perfectly comprehended my chart; and being furnished with the means, drew one of her own, very much resembling it, but with many more islands: adding also the places where we must sleep in our future progress, and those where food was to be obtained. On these points, at least, it was an emendation of the knowledge we had attained before.' There were many question books of the kind to which Ross refers; an early example,[4] dated 1739, contains the following questions and answers which are typical of the teaching which was still being given on the use of the globes in the early nineteenth century.

'*Qu.* How many sorts of lines are to be observed on the breadth of the globe?

Ans. Three sorts, *viz.* large capital lines; middling lines; and small lines.

Qu. How many capital lines are there in the breadth?

Ans. Two, the Equator, and the Zodiac.'

Greater interest in geography was probably aroused by the story books such as those written by Peter Parley. His *Tales about Canada* (1843) and his *Tales about Universal History on the Basis of Geography* (5th ed 1849) are examples of this kind of literature. Mrs Sherwood's *The Traveller* (1844) is a similar work which intersperses a few remarks on politeness among the geographical material. 'The boys', writes Mrs Sherwood, in describing the frontispiece of her book, 'are very glad to see the Traveller,— Edmund has snatched Thomas's cap from his head, and is going to wave it; might it not have been more polite, if he had taken off his own cap for the purpose?'

Another Magdalen Hall man, Joseph Guy of Tewkesbury,

matriculated at Oxford in 1828. He later wrote a textbook, *The Illustrated London Geography* (1852) which gives some idea about geographical teaching then in vogue. In his preface he says that 'there is no study children take more delight in than geography when it is properly taught', but adds that the reason why learners often recoil from the task of committing geography to memory is 'the array of difficult names necessarily presented to them at its very commencement'. To obviate this dislike he advised teachers to place maps before their pupils at the first lesson. 'They may be shown the extent of the British Empire, and the countries and islands under the sceptre of our Queen; also the settlements to which emigrants go, the spots whither convicts are transported and the missionary stations'. He writes of 'a very culpable economy in the principals of some schools, in not furnishing sufficient or proper maps for the use of their pupils'. In his view too much reliance was placed on maps that were bound up in textbooks, which could not entirely take the place of atlases or larger maps hung on the walls.

The status of geography in the Universities of Oxford and Cambridge in the middle of the nineteenth century was very low. But geography had an eminent supporter in that remarkably percipient man, the Prince Consort. In 1848 soon after he had been elected Chancellor of the University of Cambridge, he strongly deplored the fact that 'geography, modern languages, the history of art, aesthetics and other subjects' did not form courses of study for Cambridge undergraduates; he would have been equally disconcerted to discover that the position at Oxford was not much different.[5]

As the century developed a large number of books on physical geography appeared in quick succession. Mary Somerville's famous *Physical Geography* was first published in 1848; there were many flagrant copies of her admirable book and these were not always so geographical as the original.[6] Most of these textbooks dealt with physical processes only; some of them tried to relate physical geography to human life, but such attempts were infrequent. In spite of this activity a recommendation made under an Ordinance by the University Commissioners in 1857 to establish a Waynflete Professorship in Physical Geography at Magdalen College, Oxford, was later abandoned. Sir Halford Mackinder repeatedly asserted that the generalities of the old physical geography, which he regarded as a specialism of geology, were responsible for the low state of geography during the nineteenth century.[7]

John Richard Green, the historian, summed up the position of geographical education in British schools in 1880 in a book written jointly with his wife Alice Stopford Green, in the following passage:

> No drearier task can be set for the worst of criminals than that of studying a set of geographical text-books such as the children in our schools are doomed to use. Pages of 'tables', 'tables' of heights and 'tables' of areas, 'tables' of mountains and 'tables' of table-lands, 'tables' of numerals which look like arithmetical problems, but are really statements of population; these, arranged in an alphabetical order or disorder, form the only breaks in a chaotic mass of what are amusingly styled 'geographical facts', which turn out simply to be names, names of rivers and names of hills, names of counties and names of towns, a mass barely brought into grammatical shape by the needful verbs and substantives, and dotted over with isolated phrases about mining here and cotton-spinning there, which pass for Industrial Geography. Books such as these, if books they must be called, are simply appeals to the memory; they are handbooks of mnemonics, but they are in no sense handbooks of Geography.'[8]

The raising of geographical education from the slough into which it had fallen was due, in the first place, to the efforts of the Royal Geographical Society. That fact should never be forgotten. The Society, founded in 1830, was the outcome of a travellers' club, and for a long time its main and only interest lay in the promotion of exploration. It was some years before the Society took any steps to foster geographical education, but in 1866 the Society's Council began to award an annual prize of £5 to candidates at the Society of Arts examinations for their work in geography; this grant was discontinued in 1873. In 1869 the Society at the suggestion of Francis Galton,[9] offered two gold medals and two bronze medals as prizes for competitions in physical geography and in political geography at the principal Public Schools. Sixty schools took part, some of them half-heartedly, and the medals were given for the last time in 1883. In all, sixty-two medals were awarded and two schools (Dulwich and Liverpool College) carried off thirty between them. Eton obtained five and Rossall four. The Council of the Royal Geographical Society was very disappointed in the results of its educational efforts, and in 1884 it was decided to make a thorough inquiry into the teaching of geography at home and abroad. There was no need for any research to discover that the position of geography in British schools and universities was thoroughly unsatisfactory. On the other hand little was known in England about the very different state of affairs with regard to the teaching of geography in the schools and universities of foreign countries. John Scott Keltie was appointed the Society's Inspector of geographical education and he travelled abroad collecting material. His *Report*[10] dated May 1885 reveals not only the low standard of geographical teaching prevailing in England at that time, but also measures the immense progress which has been made since its publication. One English headmaster told Keltie that geography was 'little more than an effort of memory' and 'quite worthless educationally'. Keltie's most important conclusion was that no improvement could be effected in the schools unless the universities gave the subject a real place in their examinations. He thought the immediate prospect at Oxford was not so hopeful as at Cambridge; as it turned out the reverse was true. Keltie pointed out that in Germany there were twlve university chairs in geography and that geography was compulsory at all levels of school education.

In 1871 the Royal Geographical Society had appealed to Oxford and Cambridge 'not only to rescue geography from being badly taught in the schools of England, but to raise it to an even higher standard than it has yet attained'. The President's memorandum continued: 'We speak of geography not as a barren catalogue of names and facts, but as a science that ought to be taught in a liberal way, with abundant appliances of maps, models and illustrations.'[11] Again in 1874 the Council of the Society approached Oxford and Cambridge and strongly urged 'that there is no country that can less afford to dispense with geographical knowledge than England, and that, while there is no people who have a greater natural interest in it . . . there are few countries in which a high order of geographical teaching is so little encouraged'.[12] The Society urged 'the establishment of Geographical Professorships at both Universities'. Both these approaches to Oxford and Cambridge had no result. In 1886 the Society, stimulated by Keltie's *Report* tried once again. In February 1887 a delegation from the Society visited Oxford and at the end of the month it was announced that the university had decided to establish a Readership in Geography for five years. The salary was to be £300 towards which the Society was to contribute £150. Thus Oxford has the credit of being the first university to take action. Cambridge created a Lectureship in the following year, but that post was not converted into a Readership until 1898.

In July 1887, it was announced that Mr Halford J. Mackinder had been appointed to the Oxford Readership.[13] The statement added that his 'geographical lectures have attracted large audiences during the past two seasons at the chief centres of the Oxford University Extension in the north and west of England. The interests of geography as an important and definite branch of knowledge and as a necessary element in education will be safe in Mr Mackinder's hands'. When the delegates from the Society visited Oxford in February 1887, they must have been able to assure the university that there was a potential candidate for the proposed Readership who possessed the knowledge and power to make geography attractive.[14] It seems highly probable that the availability of Mackinder for the post was a determining factor in persuading the university authorities to establish the Readership.[15] Halford John Makinder was the pioneer who was destined to breathe new life into the dry bones of academic geography.

NOTES

1 This chapter is the first part of a lecture given to the annual conference of the Geographical Association in the London School of Economics on 5 January 1950, and to the Liverpool and District branch of the Association on 18 January 1951. It was printed in *Geography*, 36 (1951), 21–43 as 'Seven lamps of geography'.

2 In an essay, *Of Education*, written to Master Samuel Hartlib in 1644 in the form of a letter John Milton said: '. . . it will be reasonable for them to learn in any modern Author, the use of the Globes, and all the Maps; first with the old names and then with the new'.

3 Ross, Sir John, *Narrative of Second Voyage in Search of a North-West Passage* (1835), 261–2.

4 *An Introduction to Geography by way of Question and Answer* (1739), 4–5.

5 Winstanley, D. A., *Early Victorian Cambridge* (1940), 202; Fulford, Roger, *The Prince Consort* (1949), 199.

6 Baker, J. N. L., 'Mary Somerville and geography in England', *Geogr J*, 111 (1948), 207–22, contains a detailed account of Mary Somerville's work, and also describes the position of geographical studies in English universities in the nineteenth century.

7 *Geogr J*, 57 (1921), 377

8 Green, J. R., and Green, Alice S., *A Short Geography of the British Islands* (1880), introduction, vi–viii.

9 Freeman, T. W., *The Geographer's Craft* (1967), 22–43, for an interesting but not wholly sympathetic account of Francis Galton. The fifth edition (1872) of Galton's *Art of Travel* was re-published by David & Charles in 1971.

10 *RGS Supplementary Papers*, I (1882–5), 439–594.

11 ibid, 518.

12 ibid, 521.

13 *Proceedings of the RGS*, 9 (1887), 437.

14 H. J. Mackinder was elected a Fellow of the RGS on 22 March 1886. He was proposed by H. B. George, Fellow and Tutor of New College, Oxford, author of several books on historical geography, and seconded by Douglas W. Freshfield, later to become President of RGS in 1914–17.

15 A useful account of the position of geography in the period 1884–7 can be found in a paper by Unstead, J. F., 'H. J. Mackinder and the new geography', *Geogr J*, 113 (1949), 47–57.

Plate 9   *John Snow* (1813-58), *author of the 'water-borne' theory of cholera, one of the founders of epidemiology and the first specialist anaesthetist*

Plate 10   Henry Wentworth Acland. Crayon by George Richmond, RA, 1846

CHAPTER EIGHT

Sir Halford Mackinder
(1861–1947)

SIR HALFORD MACKINDER was born on 15 February 1861;
and it is right that we should commemorate the centenary
of the birth of this great geographer and public servant—
the second Director of the London School of Economics.[1] I was
very honoured by the invitation to deliver this lecture; and first I
must explain why I had the temerity to undertake the task. Sir
Halford was my senior by about forty years. I never worked with
him as a colleague, nor did I know him at all intimately, but I
have enjoyed certain advantages which embolden me to make
this attempt to outline his life and achievement. In 1923 I went
down from Oxford and began to teach in the University of London. I heard Sir Halford speak on several occasions during his
last two years here but I did not then meet him; he retired from
his London Professorship in 1925. However, it has so happened
that I have taught in each of the three universities in which
Mackinder held appointments—Oxford, Reading and London.
In this way I have come to learn something of the tradition which
Mackinder left behind him in these three universities. For a time
Mackinder held office in all three Thames valley towns simultaneously. It is said that he then owned three dress suits, one kept
at Oxford, one at Reading and another in London. This helped to
lighten the weight of his case when he rushed by GWR from one
official function to another.

During my years in Reading and Oxford I got to know

Mackinder personally: I met him occasionally and corresponded with him from time to time. In June 1946, at Mackinder's request, I went to see him in London; he called me to his bedroom and for over two hours he told me the story of his life. He was writing his autobiography and asked for my help in checking facts and dates. On my return that night I worked into the small hours writing down every word I could remember of his conversation. Only a few months later, early in 1947, Mackinder died; the writing of his autobiography had only just begun.

Unfortunately my knowledge of Mackinder's years in London is somewhat limited. This gap has been made good by the generosity and kindness of Mr L. M. Cantor, Lecturer in Education at the University College of North Staffordshire, who lent me the MS of his unpublished work on Mackinder. I have made considerable use of his chapter on Mackinder's life in London in preparing one section of my lecture. With this acknowledgement of my great debt to Mr Cantor I end this personal explanation.[2]

I will divide this lecture into three parts. First I will outline Mackinder's life. I will then give you a little more detail about his work in the University of London, which you can fit into the general pattern of his career. Thirdly, I will select a few of his ideas, most of them expressed over fifty years ago, but which I regard as especially applicable to our own time. In one hour it is not possible to do justice to the fertility of his mind. Just as Captain James Cook outlined the broad features of the Pacific Ocean and its lands, leaving others to fill in the detail, so Halford Mackinder sketched out many new ideas, not all geographical, for his successors to explore and develop.

Halford John Mackinder was born at Gainsborough in Lincolnshire, the eldest son of Draper Mackinder, a country doctor (Plate 12 (*a*)). He was named Halford after his maternal grandfather, Halford Wotton Hewitt of Lichfield, twice mayor of that city. The Mackinders were descended from a Scottish family, which changed its name, possibly from Macindoe, on going into exile after the rebellion of 1745. As a boy Mackinder used to hear and see the tidal wave on the Trent, known as the Eagre,[3] and he told me that this impressive sound and sight aroused his curiosity in the physical processes of nature (Plate 12 (*b*)). As he was brought up on the borders of the two counties of Lincolnshire and Nottinghamshire he soon became interested in the topography of both these large units. Mackinder sometimes spoke of George Eliot's novel *The Mill on the Floss*, published in 1860, only a few months before he was born. In this regional novel the town called 'St Oggs' was drawn from Gainsborough.

Mackinder's earliest memory of public affairs went back to a day in September 1870, when at nine years of age he had just begun to attend Gainsborough Grammar School; and took home the news, which he had learnt from a telegram on the post office door, that Napoleon III and his whole army had surrendered to the Prussians at Sedan. Mackinder died two years after Hitler's final defeat; his life had spanned the rise and fall of both the Second and Third Reich. He also remembered that there hung in his schoolroom at home in Gainsborough a picture of the famous naval engagement, that took place in March 1862, during the American Civil War, between the *Merrimac*, the first armoured ship, and the *Monitor*, the first turret ship. The era of the supremacy of the armoured ship also coincided with the span of Mackinder's life. While still at Gainsborough Grammar School he read an account of Captain Cook's voyages and began his lifelong interest in the history of travel.

He next entered Epsom College, which had been founded twenty years earlier for sons of the medical profession; his father intended him to be a doctor. He sat for scholarships at Oxford, and in 1880 Christ Church offered him a Junior Studentship in Physical Science. He devoted his main attention to biology, and in 1883 he obtained a first class in the Honour School of Natural Science. In the same year he served as President of the Oxford Union; he had spoken in a debate during his first fortnight at Oxford in October 1880 when the future Lord Curzon was President. He then read the School of Modern History in one year and obtained a second class in 1884; he was awarded the Burdett-Coutts scholarship in Geology in the same year. He read for the Bar and was called at the Inner Temple in 1886.

In 1885 Mackinder began to play an energetic part in forward-ing the Oxford University Extension scheme for adult education. It is necessary to say something about this organisation if we are to understand Mackinder's work at Reading and to some extent in London. The Oxford University Extension movement, like many other Oxford movements, was supported by men of high ideals; it originated in a desire to bring university teaching to the people of England by travelling lecturers. Most of the group of men who were identified with Oxford Extension in the eighties came from Christ Church, where Paget, afterwards Bishop of Oxford, was Dean. Mackinder and his friend Michael Ernest Sadler, of Trinity College—they had been fellow-officers of the Oxford Union—now became spearheads of the Oxford Extension movement. Sadler (later Sir Michael) had been appointed secretary of the Oxford Extension lectures committee in 1885, and one of his first acts was to invite Mackinder's help. In November the two men travelled together to Masboro in Rotherham, Sadler to lecture on 'Co-operation', Mackinder on 'Physical Geography', to an appreciative audience of 400 members of the Artisans' Co-operative Society. Mackinder was soon lecturing up and down the country on what he called 'the New Geography'.[4] He preached this gospel with all the zeal and fervour of a Wesley. Like the other Oxford missionaries of learning he was deeply convinced of the value of this effort to spread knowledge more widely among the people by extension lectures. After giving a course of lectures on 'Physical Geography' in Barnsley (Sadler's birthplace), Mackinder reported that 'in no part of England are our extension lectures more required and more appreciated than in these popu-lous districts with their very rare opportunities for higher learn-ing'. Mackinder gave 102 extension lectures in 1887-8 alone and 600 such lectures in all.[5] He travelled 30,000 miles in three years while engaged on this task. Only his strong constitution enabled him to carry out such arduous part-time work. The map (Fig 18) shows the towns in which Mackinder gave extension courses between 1885 and 1893.

The fame of Mackinder's lectures on 'the New Geography' spread far and wide, and soon came to the notice of the Royal

TOWNS IN WHICH
H.J.MACKINDER
gave courses of
OXFORD UNIVERSITY
EXTENSION LECTURES
1885-1893

Fig 18

Geographical Society. It so happened that the RGS was just then considering an approach to the ancient universities with suggestions for improving geographical studies. Mackinder was asked to write down what he meant by 'the New Geography'. In January 1887 he lectured to the RGS, by invitation, on 'the scope and methods of geography'.[6] This lecture was a remarkably mature performance for a man of only twenty-five. It was also audacious, for Mackinder vigorously attacked Major-General Sir Frederic Goldsmid for views expressed in his presidential address to Section E (Geography) of the British Association at Birmingham in the previous year. It is not surprising that during the lecture an 'Admiral, a member of the RGS Council who sat in the front row kept on muttering "damned cheek".'

In the following month Oxford decided to establish a Readership in Geography, half of the salary being provided by the RGS; in July, Mackinder was appointed to the post. Mackinder used to say he was the first Oxford Reader in Geography since the Elizabethan Richard Hakluyt, also a Christ Church man. Like Mackinder, Hakluyt had concerned himself with the politics of his day and had proved how geographical learning could benefit the State.

Mackinder was at Oxford as Reader in Geography from 1887 to 1905. He has described his first lecture at Oxford in these words: 'There was an attendance of three, one being a Don, who told me that he knew the Geography of Switzerland because he had just read Baedeker through from cover to cover; and the other two being ladies who brought their knitting, which was not usual at lectures at that time.'[7] During the first few years of his tenure he became firmly convinced that, in order to secure the serious study of geography in the schools of this country, it was essential to establish an institute from which teachers could obtain a qualification, but he did not realise that aim until 1899. In that year Mackinder became the Director at Oxford of the first British University School of Geography. The institute might well have been established in London, as Mackinder had originally envisaged.[8] W. A. S. Hewins, the first Director of LSE, had talks with J. S. Keltie, secretary of RGS, on this question in 1895–6 but there was no result. Negotiations between the RGS and the University of Oxford were more fruitful. With considerable financial assistance from the RGS the Oxford School of Geography was opened on the upper floor of the Old Ashmolean in January 1900: the first examination for the University's Diploma in Geography was held in the following year. The staff originally consisted of Mackinder, the first Director of the School, assisted by A. J. Herbertson, G. B. Grundy and H. N. Dickson (Plate 15). Only those familiar with the difficulties and hazards which attend the introduction of a new study into the curriculum of an ancient university can fully understand the labour and the qualities needed for success in such an enterprise. By 1915, ten years after Mackinder's resignation of his Oxford readership, his original aim of raising a body of qualified teachers had been achieved. Oxford alone had awarded Diplomas in Geography to seventy-seven persons and Certificates to eighty, while other universities had followed Oxford's lead.

Mackinder told me that he liked beginnings. One such 'beginning' was the foundation of the Geographical Association. On 20 May 1893 Mackinder held a meeting which really began the reform of geographical teaching in British schools. On that day he took the chair at a gathering of eleven public schoolmasters, held in the New Common room, Christ Church, at the instigation of B. B. Dickinson of Rugby. At this meeting in Oxford, Mackinder proposed from the chair 'the formation of an Association for the improvement of the status and teaching of Geography'. He can, therefore, be regarded as the co-founder of the Association; certainly he fostered its interests throughout his long life. This small 'beginning' was an important event for by 1970 the Geographical Association had a total membership of about 8,000. It was fitting that the birthplace of the Geographical Association should have been Christ Church, the college of Richard Hakluyt, Halford Mackinder and later of Percy Roxby.

In 1892, while still Reader at Oxford, Mackinder had started another great 'beginning', which led in due course to the foundation of a new university at Reading. Michael Sadler, the secretary of the extension lectures committee, had been elected Student of Christ Church in 1890. In 1885 only twenty-seven courses of lectures were provided but by 1893 Sadler was sending Oxford lecturers to nearly 400 courses in different parts of England. Sadler also introduced many innovations into extension work, such as travelling libraries for lectures, and summer schools which brought extension lecture students to the older universities in the long vacation. W. A. S. Hewins of Pembroke College was secretary of the first of these Oxford extension summer schools in 1888. Above all Sadler wanted to make the work more permanent and secure by starting colleges in provincial towns. In 1887, Benjamin

Jowett, Master of Balliol, explained how a system of perambulating lecturers should obtain continuity. 'We begin', he said, 'with a few lecturers. . . . This is the missionary stage of the enterprise. Then it becomes evident that the work would be better done if the teachers, or some of them, were always on the spot, mingling with the society of the place. We need buildings, and then comes the time of establishing a college, so that one naturally grows out of the other.'[9] The Oxford University Extension lectures had been particularly successful in Reading. In May 1892 Christ Church elected Mackinder to a Senior Studentship (Fellowship), and at the same time, presumably at the instigation of Michael Sadler, the Oxford college offered his services to the University Extension Association of Reading. The new Reading College was opened in September by the Dean of Christ Church with Mackinder as its Principal. He spent eleven years in developing the college which was in due time to receive its charter as a university. In 1896 the College of Heralds granted Reading a coat of arms which includes an engrailed cross, and a Lancaster rose on the lower part of the shield; this was derived from the arms of Christ Church, Oxford, to whose initiative the new college owed its origin. Mackinder was still Principal in 1902 when Reading became a University College. The efforts of Sadler and Mackinder, by their example, did much to promote the growth of the civic universities and university colleges in England. In 1903 in an essay on university education, Mackinder said that 'every local university must have its sphere of influence around', and he added 'these must become the focus of local patriotism'. The ultimate victory of the ideas of the university extension pioneers has come in our own day, in the sixties, with the establishment of seven new English universities. In passing I should remind you that Mackinder gave Oxford Extension lectures at both Brighton and Canterbury, towns which are now seats of universities.

It must be remembered that Mackinder was seldom able to visit Reading more than once or twice a week as he had so many other irons in the fire. In spite of this he imbued the Reading college with 'his own spirit of aspiration', to use Dr W. M. Childs' phrase. Childs was Mackinder's lieutenant and successor at Reading. This is what Childs said about Mackinder: 'Our work had gripped our imaginations. I lay it upon Mackinder. He himself was a talker, convincing and provocative. He had a way of blending dreams and hard sense, subtlety and simplicity, and he never seemed to know when he passed from one to the other. He made some opponents, as a leader in stark earnest is bound to do. He sometimes ploughed ahead leaving a wake of troubled waters, and he certainly gave the rest of us plenty to think about. Masterful, he yet made us his partners. We could always speak our minds; our criticisms were considered; sometimes they were even acted upon. But before engaging our chief in argument, it was well to be sure of one's ground.'[10]

But Mackinder was not only occupied with Oxford and Reading. He became, in his own words, 'a pluralist'. Three years after he started work as Principal at Reading he took part in another beginning, this time of the London School of Economics, which came into being in the autumn term of 1895. W. A. S. Hewins, whom Sidney Webb brought from Oxford to be the first Director of the new school, had known Mackinder when they worked together for the Oxford Extension lectures committee. Hewins, like Mackinder, had lectured to working-class audiences in the north of England. Hewins offered Mackinder the post of Lecturer in Economic Geography and for the next thirty years Mackinder was associated with the work of LSE. I shall return to this side of his career in more detail later, so that you can see it in relation to his life as a whole.

This was not all, for in 1898, when he held academic posts at Oxford, Reading and London, Mackinder decided to make an expedition to East Africa, 'because', as he said, 'at that time most people would have no use for a geographer who was not an adventurer and explorer'. He spent the long vacation of 1898 climbing in the Alps, and in the summer of 1899 he made the first ascent of Mount Kenya, no mean feat as both peaks of the mountain are over 17,000ft high.[11] 'What a beautiful mountain Kenya is,' he wrote, 'graceful but not stern, but as it seems to me with a cold feminine beauty.'

Just as Mackinder's academic geographical studies led him to be an explorer in the field, so his interest in politics, dating back to his days at the Oxford Union, led him to become a practising politician. In 1900 he contested Warwick and Leamington as a Liberal. When he got to the committee rooms he said, 'Have I got the ghost of a chance to get in?' 'No,' was the reply. 'Good,' he said. 'Then I'll stand. I'll take the literature with me.' He was defeated by the Hon Alfred Lyttelton by a majority of 831. The late Mr L. S. Amery told me that he persuaded Mackinder to become a Unionist: 'I think I helped to spoil his career. If he'd remained a Liberal he would have been elected in 1906 in the landslide and would probably have held office.' After another unsuccessful attempt to enter Parliament as a Liberal-Unionist for Hawick Burghs in March 1909, he was elected for the Camlachie division of Glasgow in January 1910, as a Unionist by the small majority of 434. He held his seat at the General Election in December of the same year by the even narrower margin of 26 and continued to represent Camlachie until he was defeated in 1922. Having entered the field of politics he gave his country service in many directions. He did not desert academic life entirely and divided his time between the House of Commons and LSE. In spite of his great ability as a lecturer, or perhaps because of it, he was not very successful in the House of Commons. Mr L. S. Amery told me that Mackinder never really got the ear of the House. During World War I he was a member of the committee which started the National War Savings movement in 1915–16; he appears to have been responsible for the idea of saving by stamps. While the voluntary system was still in force, he helped to recruit men for the Army by speeches in Scotland. In 1919–20 he went to Odessa as British High Commissioner for South Russia and made a vain attempt to restore cohesion in the White Russian forces. This task was no sinecure for a man of nearly sixty; he was knighted on his return.[12] He served as a member of Royal Commissions on Income Tax and on Awards to Inventors, both in 1919—and on Food Prices in 1925. Like Hewins, Mackinder was a staunch Imperialist; he became a member of two important Imperial committees. He was chairman of the Imperial Shipping Committee from its formation in 1920 until 1945. In

1926 he was sworn of the Privy Council; and also became chairman of the Imperial Economic Committee, a post which he held until 1931. In his retirement Mackinder served on the council of the RGS; in 1945 he was awarded the Patron's medal, the highest honour the Society can bestow. He died at his home in Dorset on 6 March 1947, at the venerable age of eighty-six.

Now let us turn back to Mackinder's thirty years at LSE (1895–1925), which overlap both the academic and the political sides of his career. Mackinder's post at LSE was at first very much a part-time occupation. He devoted his main energies to Reading and Oxford and only lectured about once a week in London. In his first session Mackinder gave a course of twenty-eight lectures to a class which 'though small was remarkable for the number of students from foreign universities', including Russians, Japanese and Americans. In 1898 the London University Act changed the University into a teaching body and the School of Economics applied for admission as a School of the University. This constitutional change took effect from the beginning of the 1900 session. The fourteen regular lecturers became teachers of the University, but as Mackinder's post was only part-time it was not until May 1902 that he became an 'appointed Teacher of the University in Economic Geography'. By this time his work had expanded considerably and his official duties were 'to give a course of not less than sixty lectures on Economic and Political Geography' and 'to give in every session not less than one course of lectures . . . on Applications of Geography to definite Economic and Political Problems . . . and to advise in the organisation of the Geographical Department of the School of Economics'.

Mackinder first began to take an active part in university affairs with his appointment in 1901 to the University's Board of Studies for Geography. The Board then consisted largely of geologists; its three geographers were Mackinder, Keltie and Chisholm. The deliberations of this body, according to Mackinder's diary, were often very heated. When the Board, in 1902, was considering the syllabus in geography for the new Matriculation examination, battle was joined between those who wanted to give it a geological slant, and those who insisted that it should be definitely geographical. Mackinder wrote in his diary on 9 May 1902: 'Attended meeting of Board of Geographical Studies at London U. Sharp division of opinion as to the new syllabus for Matriculation. Judd, Seeley and Miss Raisin wished to have Physical Geography—ie some preliminary General Science. Keltie, Chisholm, Laughton, Self, and (although he did not take part in divisions) Hewins wanted to make Regional Geography the basis —ie we wanted the syllabus to be definitely geographical. The minority was a trifle bad-tempered. Perhaps I was too uncompromising, but the principles at issue were important, and the two conceptions very distinct.' It is not surprising that Mackinder was able to write in his diary on 22 May: 'So in the end I have won.' This episode can conclude with another entry for 5 June: 'Attended the Geographical Board of Studies at South Kensington. We were more peaceful after a stormy opening. The geologists are a nuisance on a Geographical Board.'

Mackinder, with the help of Chisholm and Lyde, now worked hard to persuade the University to give Geography a larger place in its studies. Within the next few years it was included as a subject for the Intermediate in Arts and for the final BA pass degree. But Mackinder advised against the immediate establishment of honours schools of Geography at Oxford, Cambridge and London. He regarded such a move as premature and argued his case effectively in a letter to *The Times*.[13] He was probably right in his argument that there were not enough qualified teachers. The first British honours school of geography was not started until 1917, at Liverpool. The second was instituted at Aberystwyth a few months later.

Meanwhile, the School of Economics had moved to Clare Market, and in December 1903 Mackinder succeeded W. A. S. Hewins as Director. He was always highly regarded by, and on friendly terms with, the Webbs. In the summer of 1902 he spent a summer holiday with them at Little Buckland. The three of them would bicycle along the foot of the Cotswolds or in the Vale of Evesham and were sometimes accompanied by Bertrand Russell. Mackinder had already resigned his post at Reading, but held his Oxford Readership until 1905. His reign of four years as Director of LSE was a period of steady progress. The number of students more than doubled, and the financial position of the School was much improved. Mackinder, with his experience at Reading to guide him, did much to strengthen the School's connection with the University; an inter-collegiate system of attendance at lectures was started. As at Reading, Mackinder actively encouraged student social activities and amenities; he was always proud of having started the refectory. The first *Clare Market Review*, the School's student magazine, appeared in 1905.

Mackinder's appointment as Director brought in its train an increasing number of university duties. He was appointed to the Senate Academic Council and became its Vice-Chairman. By 1905 he was a member of eleven Senate committees. One Senate decision in which he played a leading part concerned the proposal to build the headquarters of the University at South Kensington. This proposal was strongly opposed by eleven members of the Senate led by Mackinder and Gregory Foster, Provost of University College. Already Mackinder had argued in the *Morning Post*[14] in favour of establishing a 'Quartier Latin' in London; he said, 'Some attempt should be made to draw the chief colleges into a single quarter of the Metropolis'. He continued with these words: 'The teaching of some of the undergraduates may perhaps be conducted in outlying institutions, but London will not feel the full influence of a great university, a civic organ formulating the trend of academic opinion and bringing it to bear effectively on the general life, until the post-graduate schools, the larger undergraduate schools, and possibly the Library of the British Museum have been placed in juxtaposition. Only then will an academic society arise and an organised body of student life, only then shall we see the rich effect of a university on a truly Imperial scale stimulated by a vast practical world around, and reacting on it.' Years later, on 30 June 1933, when the Senate House on the Bloomsbury site was opened, Mackinder described in *The Times* what had happened in 1905–6. Of his dissident party in the Senate he wrote: 'We met first in my room at the School of Economics and through that winter "the Eleven" dined together frequently and took counsel. A majority of the Senate were for accepting the bird in the hand, but our group was united in the view that the University must be in the centre of London, in a building erected for the purpose, which every cabman would know as the University, and that it should be near the British Museum, and within the belt of great colleges aligned from Bedford College to King's College.' Feeling ran high, but Mackinder said that at the stormy Senate meeting on 7 March 1906, 'We kept our front unbroken . . . and the negative therefore prevailed'. Now, as the University gets a firm grip on Bloomsbury, Mackinder's dreams are being realised at last.

Meanwhile, at the School of Economics, Mackinder worked in close harmony with Sidney Webb, who always had a high opinion of his ability. In 1919, when Webb was trying to persuade William Beveridge (later Lord Beveridge) to become Director of the School, he said that Mackinder had run the Scool 'with two fingers of one hand'. Beveridge, he added, could do the same and would have plenty of time to write. In 1906, Lord Haldane, a mutual friend of Webb and Mackinder, asked the latter to organise at the School a short course for Army Officers of the rank of Captain. The new Army Class received the nickname of Haldane's Mackindergarten. Mackinder devoted much time and energy to this course, which continued annually until 1914; it was not finally discontinued until 1932. This brought Mackinder into touch with the Regular Army, and he frequently lectured at Aldershot and Woolwich. In one of his extempore addresses at Woolwich on 'the strategical geography of the Near East'[15] he gave an eloquent definition of his subject. 'Geography', he said, 'is the imaginative understanding of the great regions of the earth's surface, the power of visualizing wide areas which may be the field of long campaigns; the power of extending what you see before you or beyond the horizon, and far beyond it, the power of looking upon the map of a large area, and carrying away in the mind, not simply a picture of the map, but a picture of the country; and, again, not merely that, but the power of seeing into it with the mind's eye, and of perceiving the interrelated facts which go to make up its geography—the relief of the land; the play of the winds upon that relief; the resulting rainfall as distributed in quantity and through the seasons; from the rainfall and relief, the system of drainage of the country; from the relief and the rainfall, the distribution of soils as the product of substance of the rocks; and from all these factors, the economic value of the country in its different parts; in other words its productivity; in other words its value for supply. You take to pieces the country and you reconstruct it in your imagination. You put, as it were, X-rays through it, rather than dissect it, so that you see the interplay of the parts.' Indeed these words are more than eloquent: they are inspired. The view of geography as a synthesis has never been better expressed.

After nearly five years as Director of the School, Mackinder resigned in October 1908. He wished not only to enter active political life but also to devote himself more fully to his geographical interests and he did not leave LSE. In the same year he became Reader in Economic Geography. When efforts were made later to change his post into a professorship he seems to have been averse to the idea; it was not until 1923 that he accepted the personal title of professor, only two years before his final retirement from academic life. Thus during the whole of the period in which he was a Member of Parliament (1910–22) Mackinder was also Reader in Economic Geography at LSE. During World War I he continued to take his full share in the work of the Geography Department and included a course on 'the geography of the war area'. It must have been a great satisfaction to him, when in 1918 the University of London started an honours school of Geography; the first examination was held in 1921. The Joint School of Geography (King's College and LSE) was established in 1921–2, when Mackinder was still head of the Department at LSE.

As a lecturer Mackinder was outstanding. It has been said that his were the only official university lectures at Oxford which regularly concluded with applause. Sir Charles Grant Robertson said that Mackinder and Oliver Lodge were the two best lecturers he had ever heard, 'for each had the gift of making what was really a complex subject not only plain even to the poorest intelligence, but starting in the hearer half a dozen courses of thought'. Sir Charles added that after hearing Mackinder, 'maps, particularly physical maps, positively glowed with illumination'. Students at LSE in the years before 1914 heard Mackinder when he was at the height of his powers as a lecturer. Some of them have described their recollections of him in those days. Dr Hilda Ormsby wrote in *The Times*[16]: 'Thousands who crowded to his lectures between 1910 and 1914 will be able to picture him still, as tall, erect, distinguished, he strode, always a few minutes late, up the gangway of the Great Hall, packed with eager and impatient students, to the rostrum, as he paused a moment or two to take in the array of maps that hung there, and then, turning to his audience, delivered in his sonorous voice, without ever a note, a perfectly argued and presented synthesis.' Lady Alice Bottomley, a student at LSE from 1904 to 1908 has written: 'To me the most inspiring lecturer was Mr Mackinder. As he talked you could see by the movements of his hands that he was visualising the natural forces at work in the area and the shapes on which they worked.' Sir Horace Wilson, the distinguished civil servant, also a student at LSE from 1904 has said: 'I can remember clearly the sweep and scope of what he said and always looked forward to the hour under him. His tall commanding figure and rich voice seemed to typify the magnitude of his subject and the depth of his knowledge of it. I can see him now in my mind's eye explaining the economic (and the politico-economic) significance of the Urals, the Alps or the Andes as if he were atop of one of them—or was indeed a part of them.' The late Professor H. J. Fleure wrote to me of Mackinder's 'splendid eloquence and large vision', and went on to say: 'I remember an occasion when he arrived fifteen minutes late for a public lecture (he came straight from the House of Commons) without a notion of what he was going to talk about. That must have happened often. His eye caught sight of a Haack map put up for some other purpose, and he went in and spoke with vigour and fervour for an hour, quite unforgettably.'

The third and last part of my lecture concerns some of Mackinder's ideas as we know them from his writings. He was a man of action but he was also a man of imagination and vision. Much of Mackinder's writing was in advance of his time; it was always marked by originality of thought and expression and by brilliant

Plate 11   Halford John Mackinder

Plate 12 (a) Elswitha Hall, Caskgate Street, Gainsborough. Birthplace of Halford J. Mackinder

Plate 12 (b)   The Trent Eagre at Gainsborough

generalisation. He invented the word 'nodality'; he first gave vogue to the term 'man-power' in its modern sense in the *National Review* in 1905;[17] during the last war he described Britain as 'a moated aerodrome'.

I will take three examples of the relevance of Mackinder's thinking to our own time; the first concerns education. One of the most recurrent themes in his writing is the unity of knowledge. Remember that he read two Oxford Honour Schools; not only Natural Science, but also an Arts subject, Modern History. A great deal has been heard lately about 'the two cultures of western society'. But nearly eighty-five years ago Mackinder described this problem to the RGS and suggested a solution in these words: 'One of the greatest of all gaps lies between the natural sciences and the study of humanity. It is the duty of the geographer to build one bridge over an abyss which in the opinion of many is upsetting the equilibrium of our culture.'[18] That was in 1887; Sir Charles Snow's Rede lecture on the two cultures is dated 1959 and does not attempt to solve the problem it postulates. In 1935 Mackinder returned to what he called 'the bridge character of geography'.[19] He then said that the function of the philosophical geographer was 'to integrate and not merely to accumulate knowledge. His aim is so to digest the results of specialised research that they become available as a contribution to the assessment of values. Only so does it seem to me that science itself can be adapted to the bridging of the gap between science and values, which gap is admitted to be a danger to civilisation'.[20] Similarly, Mackinder made some very pertinent remarks about the question of university faculties and the rigid division into Arts and Science. In 1916 he wrote: 'I have struggled for many years with the difficulties of time-tables and the interlocking of teaching. Might we not do something in the way of brigading subjects in a university under, say, a score of leading professors instead of in two faculties? There could then be many permutations of curriculum, each internally consistent with some leading idea, but each complete in that it aimed at imparting definite knowledge, inquisitive outlook, and the trained power of wielding ideas in languages.' Some of these seeds seem to have borne fruit in the University College of North Staffordshire (now Keele University) and also in the courses of study proposed by the new University of Sussex.

The second example of Mackinder's original thought is in the field of political geography. The application of geography to the problems of the State was always in Mackinder's mind. His paper on 'the geographical pivot of history', with its theory of 'the natural seats of power' is a classic.[21] It was read in 1904, only a fortnight before the Russo-Japanese way began, and at a time of intense Russophobia in Britain. Mackinder described a part of Eurasia as 'the pivot area' and later as 'the heartland'. In 1919 he expanded the article into a book, *Democratic Ideals and Reality*, in which he wrote his famous prophetic warning: 'When our statesmen [at Versailes] are in conversation with the defeated enemy, some airy cherub should whisper to them from time to time this saying:

> *Who rules East Europe commands the Heartland:*
> *Who rules the Heartland commands the World-Island:*
> *Who rules the World-Island commands the World.*'[22]

Mackinder's theory of the heartland as a natural seat of power was carefully studied in Nazi Germany between the wars,[23] but it received little attention in Great Britain and America until World War II. In 1943, when he was over eighty, Mackinder repeated his warning. He then stated that 'all things considered, the conclusion is unavoidable that if the Soviet Union emerges from the war as conqueror of Germany, she must rank as the greatest land power on the globe. Moreover, she will be the power in the strategically strongest defensive position. The Heartland is the greatest natural fortress on earth. For the first time in history it is manned by a garrison sufficient both in number and in quality'.[24] In the discussion after the 1904 paper it was said that Mackinder had made two main points. The first was that the world had become one, and the second and main point was 'the enormous importance to the world of the modern expansion of Russia'.[25] That paper deserves to be re-read now even though it was written nearly seventy years ago. However much you criticise it in detail, Mackinder's paper was, as John G. Winant said 'the first to provide us with a global concept of the world'.[26] Mackinder's writ-

ings on land power can be compared with those of Captain A. T.
Mahan (1840-1914) on the influence of sea power. It has been
suggested that modern air power has destroyed the validity of the
arguments of both authors. But Mackinder in 1919, and again in
1943, used the coming of air power to support his old thesis. In
any case he was convinced that the conquest of the air gave the
world's unity a new significance for all mankind.[27]

Thirdly, Mackinder also used geography as a means of studying
the major problems of society. He once said that 'Geography
presents regions to be philosophically viewed in all their aspects
interlocked'.[28] That view is not accepted in all British universities,
but the definition is still upheld in both Oxford and LSE; it is one
of the close and friendly links between the two schools of geography founded by Mackinder. In 1902 Mackinder applied these
views to London and to the whole region of south-eastern
England in a famous passage in his masterpiece, *Britain and the
British Seas*. It begins: 'London Bridge is the pith and cause of
London.' I will read the concluding paragraphs:

> The life of the great metropolis at the beginning of the
> twentieth century exhibits a daily throb as of a huge pulsating
> heart. Every evening half a million men are sent in quick
> streams, like corpuscles of blood in the arteries, along the
> railways and the trunk roads outward to the suburbs. Every
> morning they return, crowding into the square mile or two
> wherein the exchanges of the world are finally adjusted. Per-
> haps London Bridge is still the most thronged of the ducts
> through which the humanity of London ebbs and flows.

> In a manner all south-eastern England is a single urban
> community; for steam and electricity are changing our geo-
> graphical conceptions. A city in an economic sense is no
> longer an area covered continuously with streets and houses.
> The wives and children of the merchants, even of the more
> prosperous of the artisans, live without—beyond green fields
> —where the men only sleep and pass the Sabbath. The
> metropolis in its largest meaning includes all the counties for
> whose inhabitants London is 'Town', whose men do habitual
> business there, whose women buy there, whose morning

paper is printed there, whose standard of thought is determined there. East Anglia and the West of England possess a certain independence by virtue of their comparatively remote position, but, for various reasons, even they belong effectively to Metropolitan England. Birmingham, in Industrial England, is the nearest independent community, with its own heart-beat, with subject boroughs which call it 'Town', with its own daily newspapers guiding opinion along lines not wholly dictated from London.[29]

That was written seventy years ago, but it boldly outlines, not only the problem of the journey to work in London (there are now one and a quarter million people concerned); but also that of the pervading influence of the capital on the whole region of Metropolitan England—a region that extends as far as, and even beyond, the cities of Oxford and Cambridge, as Mackinder perceived. And these two problems are two of the most difficult that Britain has to face during the next twenty years. In that time the population of the South-Eastern region alone will grow larger by at least two million persons. I must remind you of the pressing need for more intensive geographical research into the planning of the region. Mackinder called Metropolitan England. This, the true London region, is much larger than the London and South-Eastern region as officially defined.

Mackinder left no direct descendants. Except for a Mackinder prize at the University of Reading and a Halford Mackinder laboratory in the University College of North Staffordshire, his name has not been perpetuated by association. There are no Mackinder chairs, fellowships or medals. Surely in the year of the centenary of his birth, it would be appropriate to found Mackinder Readerships or Fellowships, in the universities in which he taught, for the advancement of research into the application of geography to the problems of society and of the State. Mackinder wanted geography to enlighten the practical affairs of daily life. In his own words 'geography must underlie the strategy of peace, if you would not have it subserve the strategy of war'.[30] If it is not feasible to found new posts in applied geography, may I, with dues respect, suggest that you might perhaps consider the possibility of attaching the name of Mackinder to one of your two chairs of geography? Perhaps Oxford will fill this gap?

Mackinder will be remembered as a pioneer in thought and ideas, as well as in deeds. He was the creator of modern British geography as a university subject. He can be regarded as a founder of several of its branches, including those of historical geography and political geography; he made original contributions to geomorphology; but he steadfastly maintained his belief in the unity of the subject as a whole. His effect on the teaching of geography in schools was equally great; his school textbooks, each written with scrupulous care for the proper use of English, had great influence—over half a million copies were sold. But he was not only a geographer; he was also administrator, author, explorer, politician and public servant. With tireless industry and driving power he helped to build a great adult education movement in England; *and* to lay the foundations of both the University of Reading and the London School of Economics. Nature endowed him lavishly with many gifts. His forceful personality, powerful brain and superb constitution enabled him to follow several careers at one and the same time.

I have mentioned Captain James Cook twice already in this lecture. It may seem strange to couple Cook with Mackinder, but both these men of action possessed imaginative and versatile minds. After Cook had been murdered in Hawaii the College of Heralds granted a coat of arms to his family. The motto which the College attached to that coat of arms would serve equally well for Halford Mackinder—*nil intentatum reliquit*—he left nothing untried.[31]

NOTES

1 This chapter is the 'Mackinder Centenary Lecture' given to the London School of Economics and Political Science on 20 June 1961 and printed as a pamphlet for LSE by G. Bell in 1961.

2 In 1969 Mr L. M. Cantor was appointed Schofield Professor of Education in the University of Technology at Loughborough.

3 Mackinder was born in 1861 in Elswitha Hall (Plate 12 (*a*)), a fine Georgian building in Caskgate Street which runs parallel with, and close to, the River Trent; a tablet recording Mackinder's birth now adorns the house. Beckwith, Ian, *The Making of Modern Gainsborough* (1968) gives a history of the street plan of Gainsborough and describes the town at the time of Mackinder's childhood. In 1928 Vaughan Cornish visited Gainsborough to watch the Eagre and subsequently recorded his observations in a paper: 'On tidal waves which assume the form of a group of short waves'. *Monthly Notes of RAS Geophysical Supplement, 3* (1934), 183–90.

4 A printed syllabus of these Oxford University Extension lectures on *The New Geography* (1886) is in the Bodleian Library at Oxford. Another of Mackinder's printed syllabuses was *The World—the Battlefield of Wind, Water, and Rocks* (1886). Professor Alice Garnett, in a Presidential Address, gives a valuable account of the two pioneer lectures by Mackinder, given at Masboro in the winter of 1885. *Trans IBG*, 42 (1967), 21–2.

5 In 1887–8 alone Mackinder gave courses of extension lectures at Barnsley, Worksop, Leek, Bath, Taunton, Bridgwater, Enfield, Tunbridge Wells, Ashbourne, and Banbury.

6 The lecture and the long discussion after it are in *Proceedings of the RGS*, 9 (1887), 141–74. It was reprinted, as a pamphlet, with an introduction by E. W. Gilbert, by the RGS in 1951.

7 *Geogr J*, 57 (1921), 378. In his first year as Reader at Oxford, Mackinder lectured 'on principles of geography', 'the geography of central Europe', and 'the influence of physical features on man's movements and settlements'. He also gave a public lecture on 'natural roads and sites of towns in Southern Germany', attended, he says, 'by several prominent seniors of the university'. In the following year he gave twenty-one lectures on physical geography to a small audience, but he recorded that 'four members of the university, and one or two lady students honoured me with exemplary regularity.' His lectures with a more historical bias were always better attended.

8 Cantor, L. M., 'The Royal Geographical Society and the projected London institute of Geography 1892–1899', *Geogr J*, 128 (1962), 30–5.

9 Childs, W. M., *Making a University* (1933), 7.

10 ibid, 11.

11 The Ascent of Mount Kenya is fully described by Mackinder in *Geogr J*, 15 (1900), 453–86. A shorter but graphic account of the

ascent formed part of the centenary celebrations of RGS in 1930 (*GeogrJ*, 76 (1930), 529–34). The diaries kept by Mackinder during his expedition to East Africa in the summer of 1899 were presented by his family to the Bodleian Library in 1963.

12 See *Documents of British Foreign Policy 1919–1939*, first series, vol 3 (1949), 768–98, for Mackinder's official accounts to Lord Curzon of his visits to South Russia; and also Silverlight, John, *The Victors' Dilemma: Allied Intervention in the Russian Civil War* (1970), 353–4, for a general background.

13 *The Times*, 22 November 1904.

14 *Morning Post*, 5 November 1902.

15 *Journal of the Royal Artillery*, 39 (1912–13), 195–204, an extempore address given at Royal Artillery Institutions, Woolwich, 14 December 1911.

16 *The Times*, 17 March 1947.

17 'Man-power as a measure of national and imperial strength', *National Review*, 45 (1905), 136–43.

18 *Proceedings of the RGS*, 9 (1887), 145.

19 *Geogr J*, 86 (1935), 7

20 ibid, 10

21 This lecture and the discussion after it are in *Geogr J*, 23 (1904), 421–44. The lecture was reprinted, as a pamphlet, with an introduction by E. W. Gilbert, by the RG S in 1951.

22 Mackinder, H. J., *Democratic Ideals and Reality* (1919), 194.

23 See Appendix (pp 257–9) on *Mackinder and Haushofer* for a brief account of the influence of Mackinder's writings on Karl Haushofer, the Nazi teacher of *Geopolitik*. Rudolf Hess studied under Haushofer.

24 Mackinder, H. J., 'The round world and the winning of the peace', *Foreign Affairs*, 21 (1943), 595–605. The paper was reprinted, with a few alterations, as one chapter (pp 161–73) of Weigert, H. W., and Stefansson, V., *Compass of the World: a Symposium of Political Geography* (1944).

25 These points were made by H. Spenser Wilkinson (later Professor of Military History), and he also regretted that the empty seats at the meeting were 'not occupied by the members of the Cabinet'.

26 *Geogr J*, 103 (1944), 131.

27 There have been many postwar re-appraisals of Mackinder's views. Only a few are cited here: East, W. G., 'How strong is the heartland?', *Foreign Affairs*, 29 (1950), 78–93; Hall, A. H., 'Mackinder and the course of events', *Ann Ass Amer Geogrs*, 45 (1955), 109–26; Mills, D. R., 'The U.S.S.R. A re-appraisal of Mackinder's heartland concept'. *Scott Geogr Mag*, 72 (1956), 144–52; Hooson,

David, J. M., 'A new Soviet heartland?', *Geogr J*, 128 (1962), 19–29; Whebell, C. F. J., 'Mackinder's heartland theory in practice today,' *Geogr Mag*, 42 (1970), 630–6. East in 1950 insisted that Mackinder's 'geopolitical thinking is still relevant to the task of winning the peace', while Whebell, in 1970, said that Mackinder's views are as 'up to date as tomorrow'.

28 *Geogr J*, 86 (1935), 12.
29 Mackinder, H. J., *Britain and the British Seas* (first ed, 1902), 257–8.
30 *Scott Geogr Mag*, 47 (1931), 335.
31 The following lists other biographical accounts of H. J. Mackinder by E. W. Gilbert; 'The Right Honourable Sir Halford J. Mackinder, PC, 1861–1947', *Geogr J*, 110 (1947), 94–9, an obituary with photograph; a centenary note in *Geogr J*, 127 (1961), 27–9, with portrait by Sir William Rothenstein; *Dictionary of National Biography 1941–50* (1959), 566–7; *International Encyclopaedia of the Social Sciences* (New York, 1968), 515–17.

CHAPTER NINE

Foundations of Mackinder's Geographical Teaching

T HERE IS NO single book which summarises Mackinder's main doctrines.[1] He wrote a number of books, but the principal sources for Mackinder's views on teaching are his papers which lie scattered among many different periodicals, some of which are difficult to obtain.[2] After going through these papers I grouped some of Mackinder's ideas under seven headings as seven foundation stones of geography. I did not use the definite article because the choice of ideas is mine and I am well aware that this chapter is in no sense a comprehensive picture of all Mackinder's ideas. I have merely selected what are, in my opinion, Mackinder's most important contributions to geographical teaching.

The Region. The first foundation stone is the region. Herbertson is generally given all the credit for starting in this country the teaching of geography by regions. But the idea of the region permeates Mackinder's famous lecture to the RGS in 1887 on 'the scope and methods of geography'.[3] In that lecture he defined geography as 'the subject whose main function is to trace the interaction of man in society and so much of his environment as varies locally',[4] and he added that ' "an environment" is a natural region. The smaller the area included the greater tends to be the number of conditions uniform or nearly uniform throughout the area'.[5] The series of books planned by Mackinder and called *The Regions of the World*, of which *Britain and British Seas*, published in 1902,[6] was the first, aimed at presenting 'a picture of the physical features and conditions of a great natural region, and to trace their influence upon human societies'.[7] Those words were written by Mackinder two years before Herbertson read his notable paper on 'the major natural regions' to the Royal Geographical Society in February 1904. Mackinder, in an obituary notice of Herbertson, referred to the latter's first appointment at Oxford as follows: 'When we made him Lecturer in "Regional" Geography in the Oxford School, we used for the first time officially the word which was and, in my opinion, still is, the key to right method alike in school and university.'[8] In later years Mackinder often returned to the regional theme. In 1937 he put the matter more fully in the following words: 'the idea with which the School of Geography was founded at Oxford . . . was to bring the subject into its own again, by using "physical" geography for the analysis of "regional" geography. Our aim was to put the answer to the question "where" into the foreground, ond then to answer the question "how" in the form of "why there?" . . . the distributions of the various categories of phenomena in a given region having been separately mapped, it then, and then only became safe to proceed to correlate those distributions into a complex or synthesis of interlocked forms and interacting functions.'[9]

By their joint efforts at Oxford, Mackinder and Herbertson placed the study of regions in the very forefront of geographical work in British universities and this process was further developed by Herbertson's many pupils. The influence of what was in origin an Oxford movement has been very great in the schools as well as in the universities of the country. Ultimately, by means of regionalisation, it has even affected the administration and government of Great Britain.

The Homeland and the Commonwealth. The second foundation is the teaching of the regional geography of the British Isles, the homeland, and of the British Empire, or as it is now called the Commonwealth. Mackinder's book, *Britain and the British seas* (1902) is a classic, a work of art, and although out of date in some respects, it cannot be neglected by any serious student of geography. Mackinder's school textbook on the same subject, *Our*

Own Islands (1906), was written with equal care. Dr Ormsby has told how Mackinder wrote and re-wrote his elementary text-books with the object of giving lucidity and precision to every sentence.[10]

But Mackinder did not confine himself to the home country; he did his utmost to improve the teaching of the geography of the British Empire in all schools throughout its vast expanse. He was a member of the visual instruction committee of the Colonial Office. Under that committee's auspices he wrote *Seven Lectures on the United Kingdom* and six different editions of this book were published for use in various parts of the Empire.[11] These lectures were illustrated with carefully chosen sets of lantern slides. His eight lectures on India were produced in 1910 with the authority of the same committee. Incidentally, Mackinder must be regarded as one of the pioneers in the development of visual education. B. B. Dickinson, the founder of the Geographical Association, was the chief pioneer in the use of lantern slides and photographs in geographical teaching. In fact, Dickinson's main reason for calling the meeting which resulted in the foundation of the Association was to discuss the use of the lantern in teaching and to pool the available slides, especially those of the Empire.[12] Mackinder believed in promoting mutual knowledge of the various parts of the Empire and he regarded visual instruction by lantern slides and pictures as one of the most effective means of carrying out his aims.

Two of Mackinder's papers are concerned with the teaching of the geography of the Empire[13] while a third on the English tradition and the Empire reveals the influence of Lord Milner on Mackinder's thought.[14] In his autobiography[15] H. G. Wells tells of a small dining club called the Coefficients to which he himself, Lord Milner, Mackinder and others belonged during the period 1902–8. The club included many young Imperialists among whom L. S. Amery and Mackinder were the most prominent. As Mackinder was such a firm believer in the British Empire, it is not surprising to find that eventually he held the office of chairman of the Imperial Shipping committee for twenty-five years, and, for a time also, that of chairman of the Imperial Economic committee. It has been said that all the reports of the Imperial Shipping committee 'bore plainly the impress of Sir Halford Mackinder'.[16]

World Geography. The teaching of world geography is the third foundation stone. Although Mackinder placed the study of the region in the front of geographical thought and laid especial emphasis on the regions of Britain and her Empire, he never forgot to treat the world as a whole. 'In geography,' he said, 'no mere region is self-contained.'[17] It is clear that his own personal consciousness of the earth as a rotating globe was very strong. 'I remember', he wrote, 'once standing at sunset on the deck of a becalmed vessel off the coast of Morocco; there was not a cloud in the sky, nor a ripple on the sea as I watched the hard western horizon *rise* over the face of the sun. At that moment I *saw* the earth rotate. It was, of course, a subjective certainty inspired by my schooled mind in the absence of any competing suggestions to the contrary. Even for me I confess that the sun normally sets.'[18] Mackinder constantly emphasised that no region could be studied in isolation. 'At Oxford,' he writes, 'we had hardly settled down to the distributional or regional conception of geography, when we were compelled to adapt our minds to a further conception. *There is no complete geographical region either less than or greater than the whole of the earth's surface.*'[19] Again and again Mackinder returned to the same theme. In the same lecture he stated that 'the distribution of each category of phenomena is necessarily spherical, or in other words a closed system.' He had always emphasised the unity or rather the continuity of all the oceans and regarded the water as the great unifying factor in the world. In 1919 he said that 'the ocean was one ocean all the time, but the practical meaning of that great reality was not wholly understood until a few years ago,' and added that 'perhaps it is only now being grasped in its entirety'.[20] The conquest of the air gave the unity of the world a new and vital significance for all mankind, a fact that was fully appreciated by Mackinder. Writing in 1942 he said that 'to-day, for the first time, the habitat of each separate human being is this global Earth'.[21]

The Unity of Geography. The fourth precept of Mackinder's teaching is the unity of geography. Just as he taught that the whole world was the geographer's real parish—a world that cannot be broken absolutely and finally into regions—so he taught that geography was a unity which should not be split into fragments. If it were so divided he feared that each portion of geography would be absorbed in some other branch of learning. Mackinder always remembered and deplored what had happened in the nineteenth century when geography was deprived of every duty except that of recording mere lists of names. 'In this country,' said Mackinder, 'the geologists had captured Physical Geography, and had laid it out as a garden for themselves, while the remnant known as "General Geography", was a no-man's-land encumbered with weeds and dry bones. Before British Geography could come into its own again it was necessary to re-annex the garden and to clear and cultivate the waste.'[22] In Mackinder's view the modern rise of geomorphology and other specialisms inside the fold of geography seriously endangered the subject as a whole. 'Somewhere in geography', he wrote in 1931, 'there is a fundamental unity which eludes us. Is not our difficulty how to weld together the geological and the human aspects of the subject? . . . Is it not perhaps the lure of geomorphology which has been misleading us? I am not prepared to go quite all the way with Professor Douglas Johnson of Columbia University, who would wholly exclude geomorphology from geography, but I am ready to regard it as a secondary and not the primary factor. Geomorphology, as it has now developed, has internal coherence and a consistent philosophy, and in their hunger for these joys many of our geographers have blinded themselves to the fact that as geomorphologists they are not in the centre but on the margin of geography. I had almost said that in escaping from servitude they had robbed the Egyptians, the geologists, and had been cursed for the possession of ill-gotten goods by a generation spent in the wilderness.'[23] In the same lecture he said that 'the temptation to describe a region in terms of geological dates is besetting and insidious. It is a dangerous practice because it tends to lead the geographer away from his duty. Geology should be to the geographer what anatomy is to the artist; the subsidiary subject makes a good servant but an ill master'.[24]

This does not mean that Mackinder was not interested in physical geography. On the contrary in his first lecture to the Royal Geographical Society, in 1887, he devoted much space to a physical description of south-eastern England. In fact he regarded physical geography as the foundation of the whole subject and gave lectures on that subject from the very first years of his appointment as Reader at Oxford. His own contributions to physical geography were not insignificant. Dr H. R. Mill once referred to Professor J. W. Gregory's 'admiration of the way in which Mackinder's speculations or rather intuitions, were confirmed by his detailed study of the origin of the sea-lochs of Scotland'.[25] But to Mackinder physical geography could not stand alone; he regarded it as 'in large measure preparatory to the subsequent chapters of the subject which deal with the humane aspect of it'.[26]

Again Mackinder was a trained historian and all his work bears evidence of his abiding interest in history and of his belief that the two subjects were closely allied. Nevertheless he proclaimed his fear of geography losing its individuality in history. In 1931 he wrote as follows: 'it seems to me that there is a little danger that we should mix history with geography without seeing clearly what we are doing. Geology is a part of history, and I think we ought to remember all the time that geography proper is a description of things in the present. Geography should, as I see it, be a physiological and anatomical study rather than a study in development. As its name implies, it should be a description, with causal relations in a dynamic rather than a genetic sense.'[27] Mackinder did render considerable service to the specialism of historical geography as such, which he regarded as 'a study of the *historical present*, an idea and expression familiar to all literary folk.'[28] But while recognising that 'in the geography of to-day are undoubtedly a number of remnants of such past geographies' he asserted that those circumstances 'should not alter the whole perspective of the main subject'.[29] The fact is that he regarded both geomorphology and historical geography as specialisms and stoutly maintained his constant belief in the unity of the subject as a whole. It has been suggested that all geographical knowledge, whether physical or human, is merely a part of history. Mackinder's insistence on the living geographical present is a valuable antidote to counteract this purely genetic view of geography, now so dominant in this country.

Mackinder has told how, at the beginning of this century, in spite of all his efforts, there was 'a fresh tendency to break the subject up again, and to attach the chapters of geography to the several sciences to which they are cognate'.[30] In spite of that he 'clung all the time to the idea that there is a fundamental unity in all geographical thought, whatever the necessity of specialisation for purposes of research into process. It was for that reason, and in protest against the disintegrating thought' which he saw becoming prevalent that he laid stress on the regional aspect of the subject. But for a time he thought he had failed. 'The universities', he says, 'established more schools of geography, but at the same time published curricula containing often very little real geography. The critic was once more emboldened to deny that geography could exist as a reputable study.' Latterly Mackinder believed that the pendulum had finally swung back and that the regional idea was in the ascendant, and he attributed the fact to the work of the Geographical Association.[31] But it seems that since the end of World War II there has been another swing of the pendulum towards specialism. It behoves geomorphologists, historical geographers, quantifiers, perceptionists, and all other specialists who travel in the many pleasant by-paths away from the main road of geography to ponder carefully Mackinder's remarks on the unity of the whole subject. It is not only a fact that interesting new by-paths are constantly being found, but it is also true that the explorers of these side-tracks include some of the most influential members of the geographical profession. Many of them have strayed from the main road like lost sheep, and, wandering as they do in the outer margins of the subject, seem to have forgotten the central purpose of their journey.[32]

The Map. The map is the fifth stone. Mackinder, when writing of the ideal geographer, said that 'as a cartographer he would produce scholarly and graphic maps; as a teacher he would make maps speak'.[33] That describes exactly what he himself was able to do. As early as 1888 the wall-maps constructed for him by B. V. Darbishire for lectures at the Oxford Extension summer meeting attracted considerable attention.[34] In a letter to the author dated 24 April 1944, Sir Halford said: 'I created quite a sensation in my first lectures at Oxford in '87–'92 by illustrating my lectures with German wall-maps, without names but with broad effects of colouring and shading . . . such things were unknown in England at that time. Keltie brought them over from his mission on the continent'. In the same letter he described his method of teaching with wall-maps. 'I used to lecture *extempore* on the maps,' he wrote, 'quite as much with my hands (plastically) as my voice. Say Wales —a sweep of the flat hand to generalise the highland—a run of the finger along the edge descending to the English Plain.' In his books also Mackinder made much use of maps and diagrams on which only a few selected facts were shown, in order to make certain argument more effective. This method of using maps for teaching is now commonplace both in lectures and in books, but the fact that Mackinder is largely responsible for introducing it into this country is often forgotten. 'In the *technique* of geography,' said Mackinder, 'the central fact is the map.'[35] He went on to say that 'the essential limitation of literature is that it must make its statements in sequence to the mind's ear, whereas geography presents its map to the mind's eye and states many facts simultaneously.' On another occasion he compared 'the ordering of facts on the map as seen by the geographical eye . . . to the ordering of the features on a face as seen by the artistic eye.'[36]

Applied Geography. The application of geography to the problems of Society and of the State was always in Mackinder's mind. His original definition of geography's object, namely to trace 'the interaction of man in society and so much of his environment as varies locally',[37] shows his belief in what we now call social geography, while in 1889 he said that the 'political geographer's is the crowning chapter of geography'.[38] Several of his papers deal with political geography and of these his paper on 'the geographical pivot of history', with its theory of 'the natural seats of power', is a classic.[39] Makinder described a part of Eurasia as 'the Pivot Area', and later as 'the Heartland'. Again it was Mackinder who first gave vogue to the term man-power in assessing the strength of a nation.[40] He always laid emphasis on the importance of geography in politics: in 1931 he forecast that the years of the middle of the twentieth century will provide 'the opportunity of the geographer-statesman'.[41]

Mackinder often spoke of 'planning long before that word had been debased by common use. He believed in the exercise of foresight, the real meaning of geographical planning, in politics and he was not ashamed to link it with idealistic conceptions. 'Unless I mistake,' he wrote, 'it is the message of geography that international co-operation in any future that we need consider must be based on the federal idea. If our civilisation is not to go down in blind internecine conflict, there must be a development of world planning out of regional planning, just as regional planning has come out of town planning. The statesman of the future must know something of the geographical natural regions if he is to build for stability. The peaceful readjustment of treaties to differential growth will postulate an informed and delicate geographical judgment. In the formation of such a judgment,' he added tartly, 'geomorphology cannot play the major part.'[42]

Democratic Ideals and Reality contains several references to planning and is itself an admirable example of the application of geographical thought to politics. Mackinder was speaking of the Heartland of Eurasia, an area now being rapidly developed by the Soviet Union, and the following passage has a lively meaning for the statesmen of today: 'if only as a basis for "penetrating" this dangerous Heartland, the Oceanic peoples must strive to root ever more firmly their own organisation by localities, each locality with as complete and balanced a life of its own as circumstances may permit of. The effort must go downward through the provinces to the cities ... The country-side in which the successful leaders visibly serve the interests of their weaker brethren, must be our ideal . . . There was a time when a man addressed his "friends and neighbours". We still have our friends, but too often they are scattered over the land . . . with too many of us, in our urban and suburban civilisation, that grand old word neighbour has fallen almost into desuetude . . . Neighbourliness or fraternal duty to those who are our fellow-dwellers, is the only sure foundation of a happy citizenship. Its consequences extend upward from the city through the province to the nation and to the world.'[43]

The Philosophy of Geography. The seventh and last foundation is the philosophy of geography. Mackinder regarded geography as a philsophy and one of the humanities rather than as a science. 'To my mind,' he said, 'geography is not so much a science as a philosophy, an art, and a literature. Let me tell you my meaning by a musical analogy. There is the bass of geology; there is the tenor of meteorology, there is the alto of agriculture, botanical, and zoological studies; and there is the treble of social, economic and strategical studies. Now geography is a harmony of all four.'[44] On another occasion at Exeter in 1942, he described geography as 'an art of expression parallel to and complementary to the literary arts. It culls its data in fact from the geographical aspects of a number of sciences, just as Plato and Aristotle drew upon the physics and biology of their day. It integrates its conclusions from the human standpoint and so departs from the objectivity of science, for it ranges values alongside of measured facts. Hence "outlook" is its characteristic; it is a philosophy of Man's environment, Man himself—his body—being an element in that environment'.[45] In one of his books he explained that he regarded geography, history and religious knowledge as 'the three Outlook subjects' in which we learn where we are in space, in time, and before God'.[46] Mackinder contrasted purely physical geography with his own conception of geography in these words—'You may devote your life to the study of a purely Physical Geography, or even of some one branch of that fascinating bundle of scientific applications, and may render good service in so doing, but, as an element in the culture of a nation, only its humane crown can give broad significance to Geography. It is as mental foundations for judgment in action that geography, history and literature have their function.'[47]

Because of Mackinder's firm belief in geography as a humane subject he often discussed its general value in education as a whole. 'I do not contemplate,' he wrote, 'that teachers for secondary schools, vitally important as they are in a democratic Society, should be the only output of the University teaching of geography. In its setting as a prime element in the new humanism, geography should be brought to bear on the training of those citizens, who, outside the technical callings, have in the past found in the classical education a way to eminence in Church and State.'[48] He saw that in the air age it was more than ever 'necessary that the next generation should contain men and women with a trained power of geographical imagination. Judgment in large affairs involves just that power of seeing vast masses of detail in orderly relationship and proportion which distinguishes the ideal geographer and the ideal historian, the one in regard to space and the other in regard to time, and both as they exist on this earth's surface. A worthy School of Geography of university rank should be a school of concrete philosophy'.[49]

As Mackinder believed geography to be a humane and literary subject he was very insistent that geographers should be precise and simple in their writing and should not employ technical jargon. In a discussion after a paper by Professor W. M. Davis on 'the systematic description of land forms', Mackinder urged geographers to be very careful about the terms they used: 'if we are to induce writers on the humane aspects of geography to recognise real physical distinctions, we must seek to use such terms as will fit into literary English writing. I would plead that in the place of long foreign words we should give a precise sense to strong and, where possible, English words. It is no doubt easy, when you are talking to experts, to describe with precision by the use of technical jargon. It is a very hard thing to write a good plain English sentence. But if we adopt the easier method, and shirk the harder, then geographers are throwing away a great opportunity which lies before them. The opportunity is that of making precise the thought and the language of the practical man of affairs, whether he be politician, or writer, or merchant, or soldier.'[50] This appeal to geographers to write clear English was strongly supported by Herbertson at the same meeting, who said that 'the more I work at geography, the less I use morphological classifications and their terms' and expressed his belief that 'we must employ, for most geographical purposes, characteristic physiological descriptions rather than purely genetic morphological terms'. On one occasion Mackinder said that 'he would have the young geographer practised in the use of an almost Ruskinian, purely descriptive language, with terms drawn from the quarryman, the stonemason, the farmer, the alpine climber and the water engineer'.[51] In a sense Mackinder was a poet as well as a philosopher for he saw that geography could be as sensuous as poetry. He knew how to write geography as literature and it is sad to reflect that so little contemporary geographical writing in the English language has approached the level set by the great master of the subject.

Mackinder will be remembered as a pioneer in the art of teaching geography. He will be regarded as the chief among the little band of men who fought and won for geography its rightful place in British education, but he himself was always the first to acknowledge the co-operation of many others in achieving that triumph, of Galton and Freshfield, of Chisholm and Herbertson, and especially of his old friend H. R. Mill In Dr Mill's words, Mackinder raised geography 'from a science to a branch of philosophy and carried it forward to enlighten the practical affairs of daily life.[52]

NOTES

1 This chapter is the second part of a lecture given in 1950 to the Geographical Association under the title 'Seven Lamps of Geography: an Appreciation of the Teaching of Sir Halford Mac-

kinder' and printed in *Geography* 36 (1951), 21–43. See chapter 7, note 1.

2 A bibliography of Mackinder's printed works is given on pp 41–3 of the above article.

3 *Proc RGS*, 9 (1887), 141–60.

4 ibid, 143.

5 ibid, 156. In a paper read to the British Association at Manchester in September 1887, Mackinder described the courses of lectures which he was planning to give regularly in Oxford. Course A on the principles of geography would be largely physical but 'course B will vary in subject from year to year but will always be an analysis of a particular region'. *Proc RGS*, 9 (1887), 699.

6 Vidal de la Blache's classic regional study, *Tableau de la Géographie de la France* first appeared in 1903, a year after Mackinder's *Britain and the British Seas*.

7 Mackinder, H. J., *Britain and the British Seas* (1902), preface, vii.

8 *Geogr Teacher*, 8 (1915), 143.

9 *Proc R Phil Soc Glasgow*, 63 (1936–7), 178–9

10 *Geography*, 32 (1947), 137.

11 The preface to the West African edition (1906) of the book states that the object of the lectures 'is to give to the schoolchildren of the West African Colonies, through their eyes as well as their ears, a true and simple impression of what the United Kingdom and its people are like'.

12 *Geography*, 16 (1931), 5–6.

13 *Geogr J*, 33 (1909), 40–78; *Geog Teacher*, 6 (1911), 79–86.

14 *United Empire*, 16 (1925), 724–31.

15 Wells, H. G., *Experiment in Autobiography* (1934) 760–65. An account of the club with thinly veiled descriptions of its members is to be found in Wells, H. G., *The New Machiavelli* (2nd ed, 1911). I am indebted to the late Prof S. W. Wooldridge for this reference.

16 *The Times*, 8 March 1947.

17 *Proc R Phil Soc Glasgow*, 63 (1936–7), 180.

18 ibid, 171.

19 ibid, 179.

20 Mackinder, H. J., *Democratic Ideals and Reality* (1919), 39.

21 *Geography*, 27 (1942), 129.

22 *Geogr J*, 57 (1921), 377.

23 *Scott Geogr Mag*, 47 (1931), 323.

24 ibid, 334.

25 *Geogr J*, 69 (1927), 214; ibid, 86 (1935), 13. Professor Wooldridge stated that Mackinder's *Britain and the British Seas* 'embodies original contributions in the field [of geomorphology], notably the

chapter on the rivers of Britain'. *London Essays in Geography. Rodwell Jones Memorial Volume* (1951), 26.

26 *Geogr J*, 34 (1909), 320.

27 *Geogr J*, 78 (1931), 268.

28 ibid.

29 ibid.

30 *Report Proc Internatl Geog Cong 1928* (1930), 307

31 ibid, 308

32 A similar view was strongly expressed by Maurice Le Lannou when discussing tendencies in French geographical thought in *La Géographie Humaine* (1949).

33 *Scott Geogr Mag*, 11 (1895), 508

34 *Proc RGS*, 11 (1889), 503.

35 *Geography*, 27 (1942), 123

36 *Report Internatl Geog Congress 1928* (1930), 307.

37 *Proc RGS*, 9 (1887), 143.

38 *Scott Geog Mag*, 6 (1890), 83. It should be pointed out that Mackinder was using the term 'political geography' in the wide sense then current when it corresponded roughly to what is now called 'human geography'. In 1888 in his *Syllabus of Home Reading in Geography* Mackinder complained that 'political geography' was 'an insufficient term for the human aspects of geography'.

39 *Geogr J*, 23 (1904), 421–37. It is of interest to note that Mackinder's map of 'the natural seats of power' was shown to the RGS on 25 January 1904; and that Herbertson's map of 'the major natural regions' was exhibited to the same Society on 29 February 1904. The Oxford School of Geography was thinking in terms of *natural* units.

40 Mackinder, H. J., 'Man-power as a measure of national and imperial strength', *National Review*, 45 (1905), 136–43.

41 *Scott Geogr Mag*, 47 (1931), 335.

42 ibid, 333–4.

43 Mackinder, H. J., *Democratic Ideals and Reality* (1919), 266–7. For a recent re-assessment of this book see Gilbert, E. W., and Parker, W. H., 'Mackinder's *Democratic Ideals and Reality* after fifty years', *Geogr J*, 135 (1969), 228–231.

44 *Geogr J*, 69 (1927), 214.

45 *Geography*, 27 (1942), 129.

46 *The Teaching of Geography and History* (1914), 14.

47 *Geography*, 27 (1942), 128.

48 *Geogr J*, 86 (1935), 12.

49 *Report Internatl Geog Cong 1928* (1930), 311. An interesting but obscurely-titled article by R. P. Moss conceals the fact that half of

it is devoted to a study of Mackinder's ideas and methods. Moss argues 'that the concepts of his [Mackinder's] "New Geography" are incapable of realisation using traditional method, but that they may be successfully developed using scientific deductive method'. Moss, R. P., 'Authority and charisma: criteria of validity in geographical method', *S African Geographical Journal*, 52 (1970), 13–37.

50 *Geogr J*, 34 (1909), 321.
51 *Scott Geogr Mag*, 47 (1931), 334.
52 *Geogr J*, 86 (1935), 13.

CHAPTER TEN

Andrew John Herbertson
(1865–1915)

IT IS A great honour to be asked to give this annual lecture to the joint meeting of the three main bodies engaged in furthering geographical studies in this country.[1] You must forgive me if I begin by making some excuses for the shortcomings of my paper. First, I never met Herbertson, who died in 1915 during the early part of World War I, when I was still at school. In the second place the number of Herbertson's former colleagues, pupils and friends is now sadly small; and their memories inevitably grow somewhat dimmer after half a century. But I have consulted a number of them, and if there are any merits in this lecture they are largely due to Herbertson's friends. Thirdly, when written sources about Herbertson are consulted, it becomes clear that his obituaries, composed in wartime, are not entirely satisfactory.[2] There is no biography of any kind, and Herbertson is not in the *Dictionary of National Biography*.

It is very right that the centenary of the birth of this man, who did so much for geography, should be commemorated at a joint session of the three British geographical organisations. Herbertson was a Fellow of the Royal Geographical Society for nearly twenty-three years, and as much of his early work consisted of careful observations in the field, including those made on the summit of Ben Nevis, he followed the best traditions of the Society. Moreover, the Society published two of his most famous papers, that on the distribution of rainfall and that on natural regions. Again, although Herbertson died sixteen years before the

Institute of British Geographers was founded, its members, who are mainly academic geographers, must regard Herbertson as one of their forerunners. He was essentially a learned man, possessed of a great wealth of knowledge, who strove to secure the proper recognition of geography in the university world. But, above all, it is the Geographical Association which should honour Herbertson: for his herculean efforts to improve the standard of geographical teaching in British schools; for his work as its Honorary Secretary for fifteen years from 1900, and for starting in 1901 its periodical, then called the *Geographical Teacher*, of which he served as Honorary Editor until his death. Moreover Herbertson was the man who, in 1900, pressed and persuaded the Association to widen its membership and embrace all teachers of geography from every type of school or college. In 1915 Sir Halford Mackinder said that the 'Association and the *Teacher* owe more to Herbertson than to any other man'.[3] The Association is indeed fortunate in that a large part of its foundations were laid by such a modest, wise and unselfish man.

Herbertson was born on 11 October 1865; he died on 31 July 1915, before he had completed his fiftieth year. It is essential to remember that he died prematurely if we are to appreciate both the magnitude of his achievement and its unfinished nature. I will divide this lecture into three parts: first, an outline of Herbertson's life; then a brief summary of his geographical work; and finally I will attempt, however inadequately, to draw a picture of the man himself. Time does not allow me to include a full account of Herbertson's contributions to meteorology and climatology, nor to describe in detail his influence on teaching methods in geography: both topics demand separate papers.

Andrew John Herbertson was born at Galashiels in Selkirkshire; he was one of several brothers. His family were building contractors—the firm having been founded by his grandfather—and were responsible for much building in the town during the industrial boom of the eighties. But in 1876, when Andrew was eleven, the Herbertsons were already rich enough to buy Old Gala House, the historic manor of the Scotts in Galashiels. The house had a stone dated 1583, which the vendor Scott carried off with him. The Herbertsons renamed the house 'Beechwood', but they moved later to 'Lawyer's Brae', still in the town of Galashiels. Herbertson was therefore a 'Borderer', but one must hesitate before calling him a typical Scots Borderer as Herbertson is not an old Border name: as a native of Galashiels he can be described as a 'Braw Lad'. When Herbertson first met Robert Aitken at Oxford in 1910, he at once recognised the latter, by some trick of speech, as a fellow-Borderer. Aitken was a 'Teeri', that is a native of Hawick, in Roxburghshire, the rival town of Galashiels and its 'Braw Lads'. For six or seven years, the young Herbertson attended Galashiels Academy, a fee-paying Grammar School, dating only from 1861: none of these border towns had ancient schools. He then spent two years at the Edinburgh Institution, now called Melville College. This school has strong Border associations, and its curriculum was modern with a strong emphasis on science, mathematics and modern languages. Immediately after his schooling he served for a time with an Edinburgh firm of surveyors which, in H. O. Beckit's words, 'strengthened a bent towards practical life and business which remained with him throughout his subsequent academic career'.[4]

The second period of his life began in 1886 when he matriculated at the University of Edinburgh at what was then the unusually late age of twenty—he was very nearly twenty-one. In those days boys went to the university earlier than today. What can be called the 'undergraduate' period of his career lasted for a very long time, technically from 1886 until 1898, when he took his doctorate at Freiburg; he never took a first degree. This period of Herbertson's life can be divided into two phases of equal length, the first from 1886-92 being essentially his student years. He matriculated, that is registered, in each of the three academic years 1886-9 and again in 1891-2. In 1886-7 he attended Professor G. O. Chrystal's classes in senior mathematics and those of Professor P. G. Tait in natural philosophy: Tait's notebooks for this year record Herbertson's marks as being considerably higher than those of most of his classmates. In the spring of 1887 he was awarded the Neil Arnott Scholarship in Experimental Physics jointly with R. Turnbull, later Inspector in the Dublin Department of Agriculture. Neil Arnott scholars were under obligation to assist the Professor of Natural Philosophy in his laboratory in the ensuing summer and winter sessions. During their tenure of this scholarship Herbertson and Turnbull aided Professor Tait in his experiments on impact and helped to construct the apparatus.[5] In 1887 Herbertson again took classes in advanced mathematics and in physics; in the following year he studied geology under Professor James Geikie, as well as advanced mathematics. In 1891–2 he attended classes in advanced mathematics, natural philosophy, practical astronomy under Professor Ralph Copeland, and agriculture and rural economy under Professor Robert Wallace· There is no doubt that in the laboratories at Edinburgh he laid the foundations for his meteorological studies which were to be so valuable for geography; Professor Tait himself was interested in meteorology.

There is a period of two years (1889–91) for which it is difficult to give an exact account of Herbertson's activities. He certainly spent a semester at the University of Freiburg-im-Breisgau in 1889–90.[6] Probably it was during this period also that he first fell under the spell of that remarkable man Patrick Geddes (1854–1932), J. F. White Professor of Botany at University College, Dundee, from 1888 until 1920. Herbertson must have studied botany since he appears as Geddes' Demonstrator at Dundee in 1891–2. Geddes had a profound influence on Herbertson, perhaps more in his capacity as a social scientist than as a botanist. Geddes was in Paris working on the results of a botanical survey in Mexico when, quite by chance, he attended a lecture by Edmond Demolins at the Sorbonne on the new 'science sociale' and the ideas of Le Play. Geddes was deeply excited by this lecture: indeed he was inspired to follow Le Play's methods of observing society. In 1887 Geddes began to hold summer courses at Edinburgh and in 1892 he acquired 'Outlook Tower' for his classes. His summer schools became famous; they have been described as 'the first real summer schools in Europe'.[7] To Geddes' summer courses came Demolins and de Rousiers as well as the geographer Elisée Reclus and his brother Élie, the anthropologist. An early pupil at the summer school was Herbertson who translated Paul de

Rousiers' *American Life* in 1892; this book of 437 pages was one of Herbertson's first publications.[8] It is worth pointing out that Herbertson's wife Frances Dorothy—they married in 1893—wrote the English biography of Le Play, probably between 1897 and 1899.[9] Curiously enough this book was not published until 1950.[10] One cannot exaggerate the great influence that Geddes exercised on Herbertson's thought.

No one who reviews Herbertson's long undergraduate career can fail to be struck by his complete disregard of normal degree courses. I believe that Professor Fleure once likened Herbertson to the medieval scholars who wandered from university to university in order to sit at the feet of the most famous teachers. At Edinburgh Herbertson matriculated always as an 'Arts' student, yet he selected 'Science' subjects only. On the other hand, although he attended most of the classes necessary for the degree of BSc, established by the Senatus in 1864, he ignored chemistry which was from the first a compulsory subject for Part I. In choosing his subjects of study Herbertson pursued his natural bent with no regard for university degree regulations. The broad basis of reading and experience, which he laid as a preparation for life, was not the traditional Scottish basis in which the classics, philosophy and rhetoric (so-called), combined with English literature, also had their place. To that extent he was not a typical Scot. Equally striking is Herbertson's disregard of paper qualifications obtained by examination. It seems strange that a man of his quality should have escaped these minor distinctions so completely, yet no medals, only one prize, and perhaps two certificates appear in his record. A relative, who knew Herbertson well as a young man, said that this independent attitude was inspired by Geddes who, it was said, 'did not approve of exams'. Herbertson's record as an Edinburgh student seems to indicate a character of unusual integrity, combined with great, but less unusual powers of mind. And integrity, as it often does, expressed itself outwardly as modesty and humility; all these characteristics remained his for life. Integrity is the hallmark of his scholarship.

At the present day we are obsessed by paper qualifications, by O levels and A levels at school, by Firsts, II 1's and II 2's at the university: all these and PhDs have become the trade-union cards of the educational industry. It is very salutary for us, the present generation, to consider Herbertson's very different attitude, and its implication of the austere ideal of knowledge for its own sake. Like the famous President Eliot of Harvard, Herbertson believed that the dominant idea of education should be 'the enthusiastic study of subjects for the love of them'.

Although Herbertson had not yet graduated, the years 1892–8 can be regarded as a distinct phase in his life in which he undertook post-graduate research, studied in foreign universities and held teaching appointments at Dundee, Manchester and Edinburgh. In the summer of 1892 he acted as Demonstrator in Botany at University College, Dundee, under Geddes; and organised the course on vegetable physiology: he also gave meteorological demonstrations at evening classes attended by local gardeners and members of the nursery and seed trade. In the same year he was awarded one of the most valuable scholarships of his time; he was nominated for a 'Royal Commission for the Exhibition of 1851 Physical Science Scholarship' of £150 a year. This award was 'tenable for two years . . . limited to those branches of Science (such as Physics, Mechanics, Chemistry) the extension of which is especially important for the national industries . . . tenable in any university at home or abroad, or in some other institution to be approved by the Commissioners'. It may seem odd that Herbertson should have been given this important scholarship, as he had no degree; Professor P. G. Tait must have thought very highly of him. As a result of this award Herbertson pursued hygrometrical observations at the Ben Nevis Observatory. The period September 1892 to January 1893 he spent at the High Level Observatory on Ben Nevis, and 'owing to the limited accommodation' he undertook the ordinary observations as well as the hygrometric work. It may be of interest to this meeting to notice that Herbertson's certificate for election as a Fellow of the Royal Geographical Society is dated 10 October 1892. He was described on it as 'engaged in meteorological research' and his residence given as University Hall, Edinburgh; his proposers were H. R. Mill, J. S. Keltie and B. V. Darbishire. In the spring of 1893 he spent some time in Paris in Professor Friedel's laboratory testing the methods used on Ben Nevis. It appears that a medical prohibition prevented Herbertson from undertaking further high-level work; this may have been due to heart trouble of some kind. In August and September 1893 Herbertson carried out his experiments at Fort William Low Level Observatory, while his colleagues conducted simultaneous observations on Ben Nevis. He was then ordered by his doctor to pass the winter in the south; he proceeded to Montpellier, where he made a series of determinations at the meteorological station, and worked with some of the departments of that university. Perhaps Herbertson made some contacts with the botanist Flahault, who was actively engaged in the mapping of vegetation at this time, but it was probably Geddes who first brought Flahault's methods to Scotland.[11] At the same time as the Ben Nevis observations, Herbertson was also associated with H. R. Mill and W. S. Bruce in oceanographic studies and was responsible for a report on physical observations carried out for the Fishery Board for Scotland in 1893.

Publication of the Ben Nevis results followed in the *Journal* of the Scottish Meteorological Society and later in the *Transactions of the Royal Society of Edinburgh*. Herbertson's association with Dr Alexander Buchan in regard to Ben Nevis led to further collaboration; and ultimately culminated in 1899, in the production of the great *Atlas of Meteorology* by these two authors jointly with Dr J. G. Bartholomew. It appears that a large share of the heavy work of compilation for this atlas fell on Herbertson's shoulders during the years 1896–9. The atlas was published 'under the patronage of the Royal Geographical Society'.[12] It is at this period that Herbertson worked for his doctorate on a subject which formed a part of his labours on the atlas. It appears that he was twice in Freiburg-im-Breisgau with his wife; incidentally, she spoke German like a native. It seems that the first visit was in 1895, and the second in 1898 when he was examined for his degree. He worked at the University of Freiburg under Professor Ludwig Neumann (1854–1925) for the degree of PhD which he obtained 'multa cum laude' in September 1898. Professor Neumann told

Herbertson that the examiners had given the thesis the highest possible commendation. Its subject was 'The monthly rainfall over the land surface of the globe'; with a slightly altered title it was published by the Royal Geographical Society as an extra publication in 1901.[13] In its preparation Herbertson received assistance from Dr Hann of Graz and from many distinguished meteorologists in all parts of the world.

In 1894-6 Herbertson was Lecturer in Political and Commercial Geography at Owen's College, Manchester. His salary was paid by the Royal Geographical Society and the Manchester Geographical Society in equal shares.[14] While in Manchester he lectured on the commercial geography of Asia. It is also recorded that he gave a lecture in urban geography: 'Edinburgh: a study of geographical cause and effect', and that the slides for the occasion were lent by Professor Geddes. Some of the work at Manchester must have been very discouraging. It was said that the Education Department regulations left Herbertson 'only the poorer material to work upon, students who feel that their geographical work is a penalty for not doing better in the subject at the Queen's Scholarship examination and who do not care to devote to it more time than the bare minimum'.[15] From 1896-9 he was Lecturer in Industrial and Commercial Geography at Heriot-Watt College, Edinburgh. During this time he was an active member of the Royal Scottish Geographical Society, served on its Council and acted as editor of its *Magazine* for a short period in 1897. He continued to keep in close touch with Geddes and gave lectures in 'Outlook Tower'. He also led some experimental field classes with Dunfermline schoolchildren.

The third and last period of Herbertson's life covers the sixteen years that he spent at Oxford. Mackinder had been appointed Reader in Geography at Oxford in 1887 and for twelve years he taught there, but without any department or institute. In the early part of 1899 the University of Oxford agreed to establish a School of Geography with the co-operation and financial support of the Royal Geographical Society; it was also decided to grant a Diploma in Geography. Mackinder was to continue as Reader and to be the Director of the new School, the first British university department of Geography. His immediate task was to look for an assistant and he often told the story of how, in the spring of 1899, he persuaded Herbertson to come to Oxford. 'One day', said Mackinder, 'a rumour reached me that a young Scottish Geographer, Herbertson, a disciple of Patrick Geddes, on whom I had had my eye for my principal assistant, had been offered a professorship in New York, which he was about to accept. There and then I jumped into a train from Oxford to London, and there took the night express on to Edinburgh. In the morning I breakfasted with Bartholomew, the great cartographer, and finding that he had as high an opinion of Herbertson's geography as I had, I drove to Colinton where Herbertson had his home. After lunch, when we were both in favourable moods, I told him that he was not going to New York, but was coming to me at Oxford in the autumn—and he came'. This was not the least of Mackinder's services to British geography. 'What attracted me to him, apart from his sympathetic nature', wrote Mackinder, 'was his wide training . . . only on the historical side was his preparation weak'.[16] In June 1899 Herbertson was appointed as Assistant to the Reader and Lecturer in Regional Geography. The use of the word 'regional' in his title was significant. 'We used', said Mackinder, 'for the first time officially the word which was, and in my opinion still is, the key to right method alike in school and university'. Herbertson received a salary of £270 per annum and a share of other fees. At long last, in 1902 he obtained the degree of MA, by an Oxford decree and became a member of Wadham College. After six years, in 1905, he succeeded Mackinder as Reader on the latter's appointment as Director of the London School of Economics. In January 1910, as a recognition of his work, Oxford awarded Herbertson the personal title of Professor of Geography. In his last years Herbertson was troubled by poor health; in 1914 and 1915 the condition of his heart caused much anxiety. Lectures were still given on the upper floor of the Old Ashmolean which had to be reached by the ascent of fifty or more steps. It used to take him ten minutes to mount the stairs and he would sink into a chair, breathless and purple; and there would be further delay before he began to lecture. In the spring he was taken ill at Rome; a few

Plate 13   Percy Maude Roxby

months later he died at his country home near Chinnor on 31 July 1915 He was buried in St Cross churchyard, Oxford, not many yards from the present School of Geography. At the funeral his doctor told one of his friends that Herbertson would have been alive that day if he had not, in the early months of the 1914 war, been 'galloping about' in the early mornings in the parks. He was a member of a volunteer body (similar to the Home Guard) and turned out regularly at its pre-breakfast drills. It was the opinion of several of his contemporaries that these drills killed him and not overwork, as is sometimes stated. Just a fortnight after his death, his widow died also. She had been intimately associated with much of her husband's work, especially in the writing of school-books. The Herbertsons had two children; their son, Andrew Hunter, born in 1894, went to Balliol, and was killed in action in May 1917. Their daughter, Margaret Alice Louisa Herbertson, qualified as a medical practitioner and worked in a poor district of Liverpool for many years.

In the second part of my lecture I will discuss, very summarily, the significance of Herbertson's geographical work during his sixteen years at Oxford, the most fruitful period of his life. From 1895 onward his concern for geographical education and especially for its parlous plight in Scotland had become steadily deeper. He

Opposite:

Plate 14 (a)   The 'School of Geography', University of Liverpool from 1921 to 1965 in Abercromby Square. The houses occupied were those behind the two left-hand doors, nos 13 and 12. They were built in 1834–7. No 12 contained Roxby's 'School', which expanded into no 13 in 1953. The Department moved into the new Social Studies building in August 1965

Plate 14 (b)   The old Ashmolean building was erected in 1678–83. Dr Robert Plot was the first keeper of the Ashmolean Museum in 1683–90. Mackinder and Herbertson started the first British University School of Geography in the upper floor of the building in 1899. Roxby worked here with Herbertson. The School left in 1910 for Acland House

was not able to do anything effective to remedy the situation until Oxford gave him the very opportunity he needed—namely, to conduct a mission for the betterment of geographical teaching in schools; he performed this task with evangelistic zeal. A list of Herbertson's early published papers shows that several of them are concerned with geographical education in schools and universities.[17] During his years at Oxford, Herbertson, with Mackinder, was able to build up a body of teachers, first in the univercities and then in the schools. Between 1899 and 1905 nine students were granted the new Oxford Diploma in Geography; they included E. C. Spicer (1901), later head of the department of geography at University College, Reading, O. J. R. Howarth (1902) and Nora E. MacMunn (1904). During the ten years from 1905 in which Herbertson was Director of the Oxford School, Diplomas were granted to sixty-eight persons. These included Eva G. R. Taylor (1908), C. B. Fawcett (1912) and Blanche Hosgood (1914), each of whom became head of a department of geography in the University of London; other names in the lists are O. G. S. Crawford (1910), Charlotte Simpson (1910), W. G. Kendrew (1911) and John Bygott (1912). Although the total number of those who gained the Diploma and of those who obtained the lesser qualification of the Certificate is not large, several of them eventually exercised great influence on the development of geography in the universities. Other Oxford students who did not read geography for a qualification, but who were influenced by Herbertson during their undergraduate years, include P. M. Roxby who described Herbertson as 'my honoured master' in his Presidential address to Section E of the British Association in 1930; and A. G. Ogilvie who attended Herbertson's lectures while reading Modern History, and was later his Junior Demonstrator at Oxford from 1912 to 1914.

But both Mackinder and Herbertson saw that there was also a crying need to train far more teachers of geography for schools including the re-training of those already at work. To achieve this aim they established short but intensive courses in the long vacation. The courses owed much in method to those organised by Geddes at Edinburgh. The first Oxford courses were held in 1902 and 1904, when Mackinder was Reader; each was attended by about thirty persons. Herbertson organised five of these biennial courses between 1906 and 1914 and the number of students increased considerably; 250 in 1910 was the maximum. There is a long account of the 1910 course in *Geographical Teacher*, with full syllabuses of some of the sets of lectures, including those given by Geddes.[18] Professor R. N. Rudmose Brown, who assisted at some of these courses, has written of the Herbertsons at this time: 'Herbertson . . . the small active man, the eager encouraging teacher, restless, continually at work, never learning to play or rest; and Mrs Herbertson, busy, evergetic—even more so than Herbertson, a driving force at the summer schools, always everywhere, keeping everyone up to scratch.'[19] A large number of those who attended the vacation courses were supported by bursaries from local authorities; 39 out of 176 in 1908. In that year students undertook field work in eight separate groups; there were afternoons surveying at Marston Ferry. The lecturers included Professor J. L. Myres, J. F. Unstead, R. N. Rudmose Brown and Dr Marcel Hardy, who also gave special instruction in plant geography. Distinguished American geographers who took part in the 1908 course were Professors A. P. Brigham, W. M. Davis, C. R. Dryer, N. M. Fenneman and D. W. Johnson. This was indeed a galaxy of geomorphologists, and W. M. Davis himself led one of the field classes to study and sketch the valley forms of the Evenlode in the Cotswolds. Over 850 students attended one or other of the five summer schools directed by Herbertson; by this means alone he exerted great influence on the teaching of geography in British schools. This influence was maintained and extended by the numerous textbooks written by him and his wife, and published by the Clarendon Press, Oxford. The same Press also produced Herbertson's famous series of wall-maps. The Secretary of the Clarendon Press has kindly informed me that the Press issued a total of over 1·4 million copies of the various orange-covered geographical textbooks written by the Herbertsons. This figure includes over a quarter of a million copies of one book, *The Preliminary Geography*, first issued in 1906, and not out of print until 1950, thirty-five years after it's author's death; it is an impressive record. For comparison it can be noted that the firm of George Philip issued over half a million copies of Sir Halford Mackinder's school textbooks.

When Herbertson was compiling the Oxford wall-maps in 1914–15 a frequent visitor to the School was H. S. Hattin, a draughtsman who lived in St Michael's Street. Hattin drew and redrew these maps for Herbertson, who was meticulous in his care for accuracy. At one of their meetings there was a discussion about a map of Poland. Hattin referred to the town of L-O-D-Z, which he pronounced 'Loads'; Herbertson called it 'Woodsh'. The following conversation took place:

HATTIN: 'Woodsh'?
PROFESSOR: Yes. 'Woodsh.'
HATTIN: Which is 'Woodsh', sir?
PROFESSOR: Here [pointing to L-O-D-Z].
HATTIN: Oh· 'Loads.'
PROFESSOR: No. 'Woodsh.' It's called 'Woodsh'.
HATTIN: Loads of 'Woodsh', perhaps·
PROFESSOR: That is not funny, Mr Hattin.

The question of Polish place-names must have been much in Herbertson's mind at this time. One of the last numbers of *Geographical Teacher* which Herbertson edited included an article on this subject by Dr Ludwig Ehrlich, in which it is stated that Lodz is pronounced as 'Woodsh' (appr.).[20]

Herbertson never spared himself in his many-sided campaign to improve the quality and standing of geography in British universities and schools; to that end he dedicated his life. He postponed and sacrificed his various schemes for scientific work in order to promote geographical education. Indeed he devoted so much time and energy to the improvement of the teaching of geography by summer schools and textbooks that his own original contributions to the subject were smaller than they might otherwise have been: yet they are of great significance and it is often forgotten how relevant to modern geographical thought many of them are. Herbertson is still best remembered for his paper on 'The major natural regions', published by the Royal Geographical Society in 1905.[21] Sir Dudley Stamp made this paper the theme of his Herbertson Memorial Lecture in 1957 and it is not necessary to add to what he said then. But I should like to repeat and, if I may, support his statement 'that it would be difficult to cite any other single communication which has had such far-reaching effects in the development of our subject'.[22] Sir Dudley added that Herbertson's point of view, as expressed in the 1905 paper, 'is now fundamental to the teaching of geography in almost every country in the world', and 'is equally basic in the progress of research in a dozen different fields'. Herbertson developed and modified his opinions in later papers, a fact that is often forgotten by present critics of the idea of the region. The most valuable of his later papers, in which he includes Man in his concept of geographical regions, is 'The higher units', published in 1913.[23] In 1957 a learned essay entitled, 'The theory of geographical zones: 50 years after A. J. Herbertson' was published by Professor Willi Czajka, of the University of Göttingen.[24] This paper deserves close study, not only because it gives full credit to Herbertson for the originality of his work, but also because it discusses his philosophical opinions in relation to German authorities such as Passarge, Hettner, Köppen, von Wissmann, Troll and many others, including Thornthwaite and Unstead. Czajka starts by asserting that 'the first attempt to divide the continents according to their nature was made by A. J. Herbertson in 1905 in a paper read to the Royal Geographical Society in London'.[25] Czajka concludes that in Herbertson's attempt 'the natural regions were recognized with the help of the zonal concept'.[26] Herbertson's zonal ideas come out clearly in his paper on 'The thermal regions of the globe', also published in *Geographical Journal* in 1912—an article which still merits careful attention.[27] It was followed by a most fruitful discussion.

In his 1905 paper on 'The major natural regions' (p 309), Herbertson stated that 'a natural region should have a certain unity of configuration, climate, and vegetation'. He illustrated his paper by four clear maps. It is obvious that he was much indebted to the work of Alexander G. Supan (1847–1920), which he must have studied when he was in Germany. Herbertson acknowledged that his map of 'seasonal rainfall' was based on Supan's four

seasonal maps. In the same paper Herbertson's map of 'temperature belts' was based on sea-level isotherms. In his 1912 paper on
thermal regions he stated (p 520) that 'in all geographical work
we find that the isotherms reduced to sea-level are insufficient'.
He had previously prepared actual temperature maps, which were
revised and redrawn for his paper by his pupils Miss Rogers and
Miss E. G. R. Taylor.

After Herbertson had read his 1905 paper to the Royal Geographical Society there followed a discussion in which cold water
was poured on his ideas by a succession of speakers. It is surprising that he was not discouraged by the paper's chilly reception. At
the present day attacks on Herbertsons notions are even more
vigorous. It is sometimes argued that the regional approach to
geography is didactic, arbitrary and not sufficiently related to
practical affairs. It should be remembered that the regional concept in geography has changed and developed; Herbertson's own
views were not constant, but they were certainly neither arbitrary
nor capricious. The best answer to the charge that Herbertson's
views are of no practical value is provided by Herbertson himself.
At the very end of his 1905 paper (p 309) he said: 'It would be
difficult to exaggerate the importance of this investigation,
which seems to me a fundamental one for all who have to deal
with the study of man, or with his economic exploitation or his
proper government.' He ended by hoping that his work would
'lead to a better understanding of one part of geography and of its
practical as well as its theoretical importance'. The economic
regions devised for England, and announced in December 1964
by the Minister of Economic Affairs, were surely sufficient evidence of the practical importance of the regional approach to
geography. And Herbertson was a very practical Scot.

At the present day the subject of Biogeography is a very
fashionable branch of geography. But more than half a century
ago Herbertson used this term and realised its scientific significance. Herbertson also saw the practical importance of what was
then usually called plant and animal geography and especially of
the mapping of plant associations. He argued that such maps
should be made of the British Isles because 'questions of reforest-

ing and of changes in agricultural products are of immense importance'.[28] He believed that 'much waste of money and time might have been saved to settlers in most of our colonies' if such maps had existed.

Let me repeat what I have already said. Herbertson was essentially a practical man, and his shrewd common sense enabled him to foresee the practical value of much geographical research. As long ago as 1898 he emphasised the need for what he called 'Applied Geography'.[29] He also foresaw the potentialities of geography as a training for administrators and town-planners. He himself served in 1906–10 as a member of the Royal Commission on Canals and Waterways and was largely responsible for the fact that its Report was so well equipped with maps. In 1910 in a far-sighted Presidential Address to Section E of the British Association at Sheffield he argued that an official Hydrographical Department should be established to plan the rational use of all the water resources of Britain.[30] In the following year, at the Royal Geographical Society, Herbertson again emphasised the urgency of the question of Britian's water supply, and pressed for precise data about river discharge as well as about rainfalls. Not until fifty years later did the Water Resources Act, 1963, become law; its object was to establish river authorities and a Water Resources Board with new functions and powers for controlling the country's water for the benefit of all. In the 1910 Presidential Address Herbertson also argued that 'it ought to be possible to map the economic value of different regions at the present day', and that 'out of them might grow other maps prophetic of economic possiblities'. He then posed a question: 'Is it too much to look forward to the time when the geographical prospector, the geographer who can estimate potential geographical values, will be as common as and more reliable than the mining prospector?'[31] In 1914, with O. J. R. Howarth, Herbertson produced the six volumes of the *Oxford Survey of the British Empire*. It is known that one of Herbertson's hopes was to see a geographical colonial institute founded in Oxford; he would have been delighted by the establishment of the Institute of Colonial Studies in Oxford in 1947 (Commonwealth Studies, 1958).

Herbertson's work in 'human geography' is equally up-to-date. While Mackinder, in 1895, may have been the first to use this term in England,[32] Herbertson and his wife gave their book *Man and His Work* (1899) the sub-title *An Introduction to Human Geography*. The fact that an 8th edition was published in 1963 is a sufficient tribute to the book's lasting value. An edition in Czech was published at Prague in 1906. Herbertson is equally modern both in his desire for sound geographical description,[33] and for using the field class as a means of training geographers to use their eyes. He was much influenced by Ruskin's words: 'It is to me a standing marvel how scholars can endure, for all these centuries, to have only the name . . . of hills and rivers on their lips . . . and never one line of conception of them in their mind's sight.' Herbertson felt the need for the careful study of what he called 'our own district' if we are to understand the parts of the world which we can never see. So Oxford students for the Diploma in Geography had to write their Geographical Descriptions, not of regions, but of the Ordnance Survey New Series sheets on the one-inch scale. Herbertson himself wrote such an account of the Oxford sheet.[34] In his paper on 'the higher units' he said that 'if geography did nothing but teach a child to see, know and love his own district, it would be an inestimably valuable element in education'.[35] Herbertson certainly loved his own Tweed valley where he was born, and of which he once hoped to write, as well as his adopted Oxford countryside between the Chilterns and the Cotswolds. He once said that 'each place has its *genius loci*, of which the poet is usually the best interpreter',[36] and advised the student of the Oxford district to read Matthew Arnold's *Scholar Gipsy* and Scott's *Woodstock*. In view of Herbertson's early training it is not surprising that in 1901 he described geomorphology to O. J. R. Howarth as 'this whole great new science'.[37] The phrase remained in Howarth's memory because at the time he did not know what geomorphology meant. But it is surprising that even in the field of historical geography Herbertson's ideas can be regarded as modern. He owned a copy of E. A. Freeman's *Historical Geography of Europe* (1881) and against Freeman's very limited definition of his subject he pencilled in his own. 'Historical geography', he wrote, 'describes and interprets human distributions at any past period and the successive changes of human distributions, economic, political, and racial in the widest sense, within a defined area throughout historical time.'[38]

Praise of a man's work is worth more if it is tempered by criticism. In the preparation of this paper I read most of Herbertson's writings. If you were to do this at one fell swoop, I think most of you would be surprised, as I was, by the crudity of some of his deterministic statements. In this respect his work is dated, but this can easily be explained. Herbertson was born only six years after the publication of Charles Darwin's *The Origin of Species* (1859). Herbertson lived at a time when Darwin's ideas of evolution were omnipresent, and were being applied to every branch of learning. The applications of the doctrine of evolution to geography can involve very great dangers. It has been shown recently that these dangers can exist even in geomorphology; but at the present time they are far greater in human geography, especially if the doctrine of progress, as popularised by Herbert Spencer, is associated with the idea of evolution. An evolutionary philosophy can make human geography far simpler than it is in reality. And the evolutionary approach to human geography can also result in the false application of scientific analogies. It is not suggested that Herbertson fell into these errors but his determinism should be examined in this context.

Under what conditions did Herbertson pursue his multifarious activities in Oxford? In the Edwardian period Oxford's attitude to geography must have been somewhat lukewarm, and Herbertson's promotion to the status of Professor in 1910 was a personal triumph. When, in 1909, O. G. S. Crawford decided to give up reading 'Greats' and take the Diploma in Geography instead he went to inform his tutor in Keble. Crawford wrote: 'It was like a son telling his father he had decided to marry a barmaid. There was nothing he could do about it and he knew it. . . . Going from Greats to Geography was like leaving the parlour for the basement; one lost caste but one did see life . . . I felt at home in the new environment of maps and things of this world . . . one felt intuitively that Geography was a subject in process of formation.'[39]

Mackinder and Herbertson had started teaching in rooms in the Old Ashmolean (Plate 14 (*b*)), but these had become over-crowded and in 1910 the School, by means of a benefaction from Mr (later Sir) Abe Bailey, was enabled to move across Broad Street into Acland House, now demolished.[40] This was a former home of Ruskin's friend Sir Henry Acland (1815-1900), for many years Regius Professor of Medicine and incidentally a Fellow of the Royal Geographical Society. It was not a convenient house for a university department, as it consisted of many small rooms, only two of which could be used for lectures and classes; Herbert-son's oak-panelled study, formerly Acland's bedroom, overlooked the garden. Robert Aitken said, 'The room would have been con-demned offhand as poky by any modern education expert, but there was a flavour about this setting which I cannot describe.' Crawford seems to have carried out the move to Acland House across the road single-handed in his own spare time; he bought a carpet for his own office at his own expense. He said 'this extravagance grated upon the Scotch austerity of Herbertson' and that he 'received a long letter from Herbertson upbraiding me for it'. Herbertson at once realized that his reproof had been too severe and made handsome amends.[41] He was in Crawford's rooms almost before the letter had been read. The Herbertsons lived in various houses in North Oxford until in 1912 they finally moved to 9 Fyfield Road. At the same time they obtained for their health's sake a country home, 'Two Shires Yew', at Sprig's Alley in the Chilterns, above Chinnor, and over 750ft above sea-level. Herbertson was an early motorist, for in 1912 he acquired a Rover car, primarily for the journeys to and from Chinnor.

During Herbertson's years in charge of the Oxford School of Geography, relations with the RGS were close. These contacts were strengthened when Lord Curzon, who had been a Fellow of the RGS since 1888, was elected Chancellor of the University of Oxford in 1907. Soon after Lord Curzon's return from India in 1905, after his seven years as Viceroy, he was re-elected to the Council of the RGS. Herbertson was Director of the Oxford School of Geography when on 2 November 1907 Lord Curzon, now Chancellor of the University, delivered his famous Romanes lecture on *Frontiers* in the Sheldonian theatre. In this lecture Lord Curzon said: 'modern works on geography realise with increasing seriousness the significance of political geography; and here in this University so responsive to the spirit of the age, where I rejoice to think that a School of Geography has recently been founded, it is not likely to escape attention.' In the same month that this lecture was delivered Herbertson presented two memoranda to the Chancellor, one on the requirements of the School of Geography, and the other on the ways in which the School of Geography could be helped. Lord Curzon served as President of the RGS in 1911–14 and obviously took great interest in the fortunes of the Oxford School of Geography; his visits to the School in Acland House, still remembered by former students, must have raised its prestige in the University.

I now turn to the third and last part of my lecture. What was Herbertson like in appearance and in character? The Herbertson of 1910, when he was 45, was described as 'rather heavily built, just tall enough not to be stocky, and still sandy-coloured, without a trace of ageing'. Another writes of 'his rather stocky figure, sandy hair and full face with its wonderful broad brow'. A former clerk in the School of Geography's office also recalls him as he was in 1910: 'Although he frightened me somewhat, he was, in fact a very kindly man and his somewhat stern expression would frequently change suddenly and he would smile benignly at one and speak in a gently soothing voice. . . . He had a florid complexion and wore pince-nez spectacles which were often fastened to a chain clipped to his coat lapel.' A former student writes: 'His best features were his blue eyes, half-hidden behind gold-rimmed spectacles; and his fresh colouring.'

Opinions differ as to his ability as a lecturer. He was described by Professor E. G. R. Taylor as a 'clear but undramatic lecturer'. O. J. R. Howarth has said that on the platform Herbertson's 'enthusiasm was in a measure concealed by a rather flat delivery'.[42] One former student writes: 'His voice was not very resonant and would not have carried in a large hall . . . it was quite adequate in the small rooms of Acland House and in the room on the first floor of the Old Ashmolean where most of our lectures took place. It gave him pleasure to know that the *Oxford Dictionary* was being compiled on the floor below. He was not an arresting lecturer, for he had no dramatic gifts of speech or manner, but students found him clear and forceful, and sometimes eolquent and inspiring.' His delivery was 'matter-of-fact' said one student; while another said that 'as a lecturer Herbertson was as dry as dust . . . but extremely interesting on the cycle rides he used to take us', and that from these field classes he, the student, obtained an interest in scenery for life. Others including A. G. Ogilvie and J. F. Unstead have paid tribute to the value of Herbertson's field classes by cycle. It is clear that Mackinder and Herbertson were indeed complementary to one another. Mackinder was the brilliant and inspiring lecturer; while Herbertson was the patient, careful and lucid tutor with a flair for practical work in the field. Oxford was fortunate to have the services of both these great pioneers.

O. G. S. Crawford, Herbertson's Junior Demonstrator from 1910 to the end of 1911, wrote of his chief: 'We all liked him for he was kindly, patient and genuinely interested in his pupils, and as a tutor he was excellent. But as a lecturer he was uninspiring at best, and at times far worse; it was sometimes quite embarrassing to have to listen to his halting delivery, as of an utterly tired-out man, which he was.'[43] Crawford adds that Herbertson 'drove himself too hard'. Crawford's opinion of Herbertson as a lecturer is confirmed by D. H. Lawrence who, when training to be a teacher at Nottingham, wrote on 11 March 1909:[44] 'We had a lecture on Geography last week by Dr Herbertson—a very great gun from Oxford and he bored me excruciatingly.' The records at Oxford show that Herbertson did the lion's share of the lecturing in the School. Whatever he was like as a lecturer, there is no doubt that Herbertson was supremely successful as an Oxford tutor and took a deep personal interest in each of his pupils. One of them has given the following account of tutorials with him: 'I would sit down at the side of his desk and he would appear to listen attentively to what was read. After hearing a few pages he would say: "Thank you. I see what you are getting at. You have taken a lot of trouble." Then he would develop the subject himself, refer one to other sources of information, throw out ideas like sparks, and relate them to their geographical background. His mind had no watertight compartments but could select whatever was relevant. At last pausing for breath he would say apologetically, his eyes twinkling—"I've given you rather a lot to think about." I never finished reading an essay, but certainly I never went away without being enriched and stimulated.'

I have already spoken of Herbertson's intellectual integrity: he was diffident and modest and always pushed others forward rather than himself. He was ever ready to make his own knowledge available to all and he took little or no credit for his own original ideas. I will read part of E. F. Elton's 1915 tribute to Herbertson: 'I desire to say a word of the immense kindliness of the man. No trouble was too great for him to take with a pupil who was in earnest . . . he was always scheming to help one or another friend or pupil; he was always planning for "the School"; I do not believe he ever gave a single thought to advancing his own interests. Again and again have I known him sacrifice them for the sake of friend or School—not unconsciously, he was too acute for that, but without hesitation as a matter of course; it was for him the natural thing to do. I have rarely known a man so loveable, never one so unselfish.'[45] That is a worthy epitaph for Herbertson the man. In conclusion let his own words provide a fitting memorial for Herbertson the geographer.

That distinguished American historian of geography, Professor Richard Hartshorne of Wisconsin, has asserted[46] that Herbertson's posthumous paper on 'Regional environment, heredity and consciousness' [47] is 'seldom if ever mentioned by British geographers' in spite of its pertinence to current discussion. Now it is true that the paper is rarely quoted in print in this country, but that does not mean that it is not read. I was delighted recently to see that one sentence from this paper is still quoted in the current pamphlet about the Department of Geography of the University of Birmingham. It is important to remember that this posthumous paper is a clear statement of Herbertson's philosophy of geography as it had become at the end of his life; his views had changed and developed.[48] For that reason alone I think it would be fitting to conclude this lecture by reading a long passage from Herbertson's last paper, which death prevented him from reading to any audience. I must warn you that one sentence, here printed in italics, is written in such forceful language that some of you may flinch. But I am convinced that his words are true, and perhaps even more applicable today than when they were first written more than half a century ago.

Environment is not merely the physical circumstances among which we live, important though those are. It is found to be more complex and more subtle the more we examine it. There is a mental and spiritual environment as well as a material one. It is almost impossible to group precisely the ideas of a community into those which are the outcome of environmental contact, and those which are due to social inheritance.

Environment is not constant, but changes, even physically, as when a new drainage or irrigation or railway system is constructed. Social tradition is not constant. In fact, heredity and environment are very convenient ideas for analysis. Abstract either element from the whole—and it is less than the whole—and the whole cannot be understood.

It is no doubt difficult for us, accustomed to these dissections, to understand that the living whole, while made up of parts with different structures and functions, is no longer the living whole when it is so dissected, but something dead and incomplete. *The separation of the whole into man and his environment is such a murderous act.* There are no men apart from their environment. There is a whole for which we have no name, unless it is a country, of which men are a part. We cannot consider men apart from the rest of the country, nor an inhabited country apart from its inhabitants without abstracting an essential part of the whole. It is like studying a human being without his nervous system, and his nervous system apart from the rest of him. It may be a useful form of analysis at a particular stage of our investigation, but it is inadequate and misleading until we have once more considered the complete man.

So it is with a country, a region, a district—whatsoever

name we care to give it. In its present form and activities man is an essential element of it, and man cannot be considered apart from the rest of it without limiting our study to something less than the whole. The analogy with an individual man is no doubt useful. There are specialists skilled in the knowledge of healthy and diseased conditions of the bones, the muscles, the digestive system, the eye, the brain, and so on; but the wise physician must know something of all of these, and as a rule this conception of the whole man is more complete and truer than that of the specialist. So with the higher natural whole or region, tinker, tailor, soldier, sailor and the thousand and one necessary specialists do not replace the wise man who knows the countryside thoroughly, and thinks of it and loves it naturally as a whole.

I will repeat those last phrases of Herbertson's: *'the wise man who knows the countryside thoroughly, and thinks of it and loves it naturally as a whole'*. Andrew Herbertson was just such a man.[49]

NOTES

1 This chapter was delivered as the Herbertson Centenary Lecture at the house of the RGS in London on 1 January 1965, to a joint meeting of the Geographical Association, the Royal Geographical Society and the Institute of British Geographers. It was repeated in Oxford on 24 February 1965 to a joint meeting of the Herbertson Society with the Oxford branch of the Geographical Association. The lecture was published in *Geography* 50 (1965), 313–31, as part of an 'A. J. Herbertson Centenary Special Issue'.

2 *Geogr J*, 46 (1915), 319–20, by Beckit, H. O.; *Scott Geog J*, 31 (1915), 486–90, by Cossar, J.; *Geogr Teacher*, 8 (1915), 143–6, by Mackinder, H. J., McMunn, N. E., and Elton, E. F.; *J of the Scott Met Soc*, 17 (1915), 34–6, by Watt, A. G. R. Crone, in his paper 'British geography in the twentieth century', *Geogr J*, 130 (1964), gives a valuable assessment of Herbertson's place in British Geography, and of his developing thought.

3   *Geogr Teacher*, 8 (1915), 143.

4   *Geogr J*, 46 (1915), 320.

5   See *Scott Geog Mag*, 31 (1915), 486; and Knott, C. G., *Life and Scientific Work of Peter Guthrie Tait* (1911), 88 and 96.

6   At Freiburg Herbertson attended courses in mathematical geography, oceanography, meteorology, survey of continents and practical uses of geography. He also attended lectures on physics, mathematics and botany. Professor Ludwig Neumann (1854–1925) then held the chair of geography at Freiburg. He was an expert on the Black Forest and a keen student of population geography, but it appears that he was also a mathematician with a lively interest in meteorology and climatology.

7   Mairet, Philip, *Pioneer of Sociology: the Life and Letters of Patrick Geddes* (1957), 63.

8   Herbertson, A. J., *American Life* (1892): a translation of de Rousiers, Paul, *La Vie Américaine* (Paris, 1892).

9   Mrs F. D. Herbertson's maiden name was Richardson; she was a goddaughter of the famous Miss Beale of Cheltenham. She was a graduate of the University of London.

10 Herbertson, F. [D.], *The Life of Frédéric Le Play*, ed by Branford, G. Victor, and Farquharson, A. (Ledbury: Le Play House, 1950). See also Brooke, M. Z., *Le Play: Engineer and Social Scientist. The Life and work of Frédéric Le Play* (1970).

11 Herbertson wrote a paper in *Scott Geog Mag*, 13 (1897), 537–41, on 'the mapping of plant associations', which is largely a review of Flahault's work.

12 *Atlas of Meteorology*, prepared by Bartholomew, J. G., and Herbertson, A. J., and edited by Buchan, Alexander (1899).

13 Herbertson, A. J., *The Distribution of Rainfall over the Land* (1901)

14 *Geogr J* 120 (1954), 118–19

15 *Scott Geog Mag*, 12 (1896), 423.

16 *Geographical Teacher*, 8 (1915), 144.

17 See, for example, 'the parlous plight of geography in Scottish education', *Scott Geog Mag*, 14 (1898), 81–8.

18 *Geographical Teacher*, 5 (1910), 337–53.

19 *Geography*, 23 (1948), 106.

20 *Geographical Teacher*, 8 (1915), 13.

21 Herbertson, A. J., 'the major natural regions: an essay in systematic geography', *Geogr J*, 25 (1905), 300–12.

22 Stamp, L. Dudley, *Geography*, 42 (1957), 201.

23 Herbertson, A. J., 'the higher units', *Scientia*, 14 (Bologna, 1913), 199–212.

Plate 15   The staff of the Oxford School of Geography with the first candidates for the Diploma in Geography, outside the Old Ashmolean in June 1901.

(Back row from left to right); Dr H. N. Dickson, Lecturer in Physical Geography, E. C. Spicer, W. Stanford, W. Bisiker, Dr A.J. Herbertson, Lecturer in Regional Geography.

(Front row): Miss Joan B. Reynolds, H.J. Mackinder, Reader in Geography

Plate 16   Vaughan Cornish

24 Czajka, Willi, 'Die geographische Zonenlehre: 50 Jahre nach A. J. Herbertson's "the major natural regions",' *Geographisches Taschenbuch* (Wiesbaden, 1956–7), 410–29

25 ibid, 410.

26 ibid, 421.

27 Herbertson, A. J., 'the thermal regions of the globe,' *Geogr J*, 40 (1912), 518–32.

28 Herbertson, J. A., 'the mapping of plant associations', *Scott Geog Mag*, 12 (1897), 537–41.

29 Herbertson, A. J., 'Report on the teaching of applied geography', *Journal Manchester Geog Soc*, 14 (1898), 264–85; 'the teaching of applied geography,' *Journal Manchester Geog Soc*, 15 (1899), 58–9; Wise, M. J., *Geog Mag*, 37 (1964–5), 124.

30 Herbertson, A. J., 'geography and some of its present needs', *Geogr J*, 36 (1910), 475. In the following year Herbertson again argued that a Government department was necessary to take up the problem of water-supply. In a speech at the Royal Geographical Society he stated that because more and more water was needed for 'industrial, domestic, as well as for agricultural purposes, it was very necessary to have precise data about river discharge as well as about rainfalls'. *Geogr J*, 38 (1911), 304.

31 *Geogr J*, 36 (1910), 477.

32 Mackinder, H. J., *Geogr J*, 6 (1895), 375, writes of 'the facts or human geography'.

33 From 1901 to 1903 F. D. Herbertson and A. J. Herbertson produced six volumes of *Descriptive Geographies from Original Sources*, each devoted to a continent.

34 Herbertson, A. J., 'on the one-inch Ordnance Survey map with special reference to the Oxford sheet', *Geogr Teacher*, 1 (1902), 150–66. This is an account of Ordnance Survey New Series Sheet 236 (Oxford).

35 Herbertson, A. J., *Scientia*, 14 (1913), 211.

36 *Geographical Teacher*, 1 (1902), 166.

37 *Scott Geogr Mag*, 67 (1951), 155.

38 Gilbert, E. W., 'what is historical geography?', *Scott Geogr Mag*, 68 (1932), 132.

39 Crawford, O. G. S., *Said and Done* (1953), 42–5.

40 Acland House was really 39, 40 and 41 Broad Street, but for the sake of convenience was known as 40.

41 Crawford, op cit, 66.

42 Howarth, O. J. R., 'the centenary of Section E (Geography) in the British Association', *Scott Geog Mag*, 67 (1951), 158.

43 Crawford, op cit, 45

44 *Lawrence in Love: Letters to Louie Burrows*, ed, Boulton, J. T. (1968), 19.
45 Elton, E. F., *Geographical Teacher*, 8 (1915), 146.
46 Hartshorne, Richard, *Perspective on the Nature of Geograpy* (1960), 5.
47 Herbertson, A. J., 'Regional environment, heredity and consciousness', *Geographical Teacher*, 8 (1915), 147–53.
48 O. J. R. Howarth wrote of Herbertson's growing interest in human geography in the last years of his life in these words: 'Herbertson planned a work on human geography; indeed I saw a fragmentary proof of the beginning of it, heavily corrected, as death compelled him to leave it.' *Scott Geogr Mag*, 67 (1951), 156.
49 A complete bibliography of all Herbertson's published works, compiled by L. J. Jay, is in *Geography*, 50 (1965), 364–70.

CHAPTER ELEVEN

Percy Maude Roxby (1880–1947)

PERCY MAUDE ROXBY was born on 21 November 1880 at Buckden in Huntingdonshire (Plate 13). His father, a man of great ability and charm, was the evangelical vicar of Buckden; his uncle, on his father's side, was the evangelical vicar of Cheltenham. It is important to notice not only the clerical background of his upbringing, but also the fact that he grew up in the English countryside before the coming of the motor-car, but after the arrival of the bicycle. These two observations illuminate a good deal of his later life. Roxby's mother was a sister of Sir James Thomas Knowles (1831–1908), first an architect and later the founder-editor of the *Nineteenth Century*. Gladstone, Tennyson, Huxley and Manning were all contributors to this famous quarterly. Roxby, when a young man, met several important Victorian personages in his uncle's house. Nor was Knowles the only figure in the intellectual world to whom Roxby was related. Others were Walter Lord of Durham University, and Maurice Henry Hewlett (1861–1923), novelist and poet. Hewlett's mother was another sister of J. T. Knowles.[1]

The Roxby family at Buckden was a large one. One of the brothers became an early airman. Another married a Linton, a cousin of Sir John Linton Myres (1869–1954), who knew Roxby from childhood, and who at one time was Professor of Greek in the University of Liverpool. Roxby went to Bromsgrove School; his form master was the Rev E. C. Owen, later Headmaster of St Peter's School, York. Roxby's father died just about the time he won an open scholarship in History at Christ Church, Oxford. After leaving Buckden his mother maintained a home for the family, first at Turvey in Bedfordshire, then at Toddington in the same county, and later at Ockley in Surrey. Roxby gained rather than lost by living in his formative years in villages. Some of Roxby's earlier geographical interests and papers were largely concerned with rural England and its historical geography.[2] Professor Wilfred Smith in his inaugural lecture to the University of Liverpool in 1951 remarked that Roxby as a young man 'wrote a number of papers on the historical and agricultural geography of England which, if he had developed them, would have made him the master of those fields, but he left them for others to cultivate once the East cast its spell over him'.[3]

At Bromsgrove the Headmaster in Roxby's time was Herbert Millington, a pupil of Thring, the famous Head of Uppingham, and a classical scholar. Roxby did not shine in the classics; a friend of Roxby's at school remembered the Headmaster calling Roxby, then a tall lanky youth, 'you incorrigible lamp-post.' When Roxby's Oxford scholarship in history was announced the Headmaster had to make an effort to realise that Roxby had brought as much distinction to the school as his classical boys. Roxby always regarded himself as fortunate to have been taught by E. C. Owen. The latter's father, S. J. Owen, was Reader in Indian History at Oxford in 1878 and Student of Christ Church. When he came to Bromsgrove to visit his son at weekends he used to meet the young Roxby.

Roxby left Bromsgrove and went up to Oxford in a by-term (Trinity Term 1899), expecting three of his Bromsgrove contemporaries to join him a term later in October. As an undergraduate at the very beginning of the new century Roxby became devoted to the practice of going on long tours by bicycle. So at the end of his first term at Oxford he arranged with two of his old schoolfriends at Bromsgrove (both classicists) to make a tour by bicycle during his first long vacation. One of the two friends was John L. Humphreys, later to become Governor of North Borneo; like Roxby he died in China when he was on local leave. The other was Charles F. H. Soulby, later a Canon of the Anglican Cathedral at Liverpool. Roxby and Soulby, both working in Liverpool, maintained a lifelong friendship. On this first tour they travelled

from Lincoln to Yorkshire by way of Bawtry, Doncaster, Ponte-
fract, Tadcaster, York, Boroughbridge, York, Market Weighton,
Beverley, Hornsea, Hull and back to Lincoln. This trip was such a
success that the three friends agreed to make a similar journey
together during the next long vacation. Humphreys dropped out,
but Soulby and Roxby, once or twice a year during the next seven
years, went together on bicycles exploring the country. They
began with a journey through East Anglia on lines suggested by
a reading of W. A. Dutt's *East Anglia* (1901) in the 'Highways
and Byways' series. They met at Bury St Edmunds and travelling
by Stowmarket, Framlingham, Parham, Saxmundham, Dunwich,
Southwold, Lowestoft, Fritton Duck Decoy, Great Yarmouth,
Norwich, Blickling, Cromer, Weybourne, Cley, Wells, Walsingham, East Barsham, Raynham, Castle Acre, Swaffham, and
Downham Market. The two friends decided to spend some weeks
of the next Easter vacation at the Barne Arms, Dunwich, and here
Roxby completed his Prize essay on Henry Grattan. Eight hours a
day was the average time spent on reading and writing in these
vacations. The Dunwich holiday was marked by a fine walk along
the coast to Aldeburgh and back, the Aldeburgh of Crabbe and
Fitzgerald. The holiday ended with a journey to Ipswich and on
to Cambridge. The countryside connected with Gainsborough
and Constable was explored; Lavenham and Long Melford were
visited. The two friends came to East Anglia again on several
occasions. By this time Roxby the historian was becoming Roxby
the geographer.

These bicycle rides with Soulby eventually led to Roxby's study
of the *pays* of East Anglia and finally to his brilliant chapter on
the region in A. G. Ogilvie's *Great Britain* (1928).[4] This beautifully written essay is still valuable, although the economic
references are dated. The influence of Vidal de la Blache's ideas
on Roxby is very apparent. Roxby had entered Christ Church in
1899, where his history tutor was Arthur Hassall. In 1902 he was
awarded the Gladstone Memorial Prize for his essay on Henry
Grattan, being the first holder of this university prize, established
in 1901. In 1903 he obtained a first class in the Honour School of
Modern History. Also placed in the same first class with him were

H. H. E. Craster, later to become Bodley's Librarian, F. M. Powicke of Balliol, afterwards Regius Professor of Modern History, and the Hon E. F. L. Wood, also of Christ Church, afterwards Lord Halifax. Roxby took his BA in 1903, but he never proceeded to the degree of MA. It is remembered that on great academic occasions at Liverpool Roxby used to walk in procession wearing a tattered and too-short gown with a faded moth-eaten Oxford BA hood. He appears to have attended some of Mackinder's lectures on historical geography while he was reading history. In the year 1903–4 Roxby attended courses in the School of Geography. He was then a candidate for the Teacher's Diploma and he proposed also to compete for the University of Oxford's Geographical Scholarship in September 1904. In Michaelmas term of 1903 he attended Mackinder's course on the historical geography of Europe, as well as Herbertson's courses on the British Isles and on types of landforms. He had tutorials with Herbertson in Hilary and Trinity terms of 1904. In the same terms he heard Mackinder on the historical geography of the British Isles and Herbertson on the regional geography of America. He then attended Herbertson's vacation course for teachers of geography in August 1904, where he met Unstead and formed a friendship which lasted forty years.[5] The two men discussed Herbertson's views on regions and developed opinions that differed somewhat from each other and from those held by Herbertson. Mackinder lectured at this course on 'the relation of geography and history' and Herbertson on 'the geography of south-east England with special reference to the Oxford district.' Roxby did not take the examination for the Diploma in Geography and he was not in residence in Michaelmas term 1904. The School of Geography minute book records that on 13 October 1904 'the Reader, H. J. Mackinder, reported that Mr P. M. Roxby, BA, of Christ Church, who had worked at the School during the past year and had attended the vacation course had been appointed assistant lecturer in Geography at the University of Liverpool'. During his one year at the Oxford School of Geography Herbertson made a very deep impression on Roxby. He regarded Herbertson with sincere affection and Herbertson took the place of Owen as Roxby's main personal inspiration. When Roxby gave his presidential address to Section E of the British Association for the Advancement of Science at Bristol in 1930, he referred to Herbertson as his 'honoured master'. The University of Liverpool obtained its charter in 1904 and Roxby went there in October of that year as an assistant lecturer in Geography, but in the Department of Economics under Professor E. C. K. (later Sir Edward) Gonner. Roxby was therefore only half-trained as a geographer when he went to work under Gonner, an economist. Sir John Myres once remarked that Roxby 'was snatched away half-trained' from Oxford. Gonner advised Roxby to spend several afternoons at the docks at Liverpool as the best introduction to his life in the city and to his career as a teaching geographer. Roxby thus began his connection with the University of Liverpool which lasted for forty years, until he left for China in 1944. He became Lecturer in Regional Geography in 1906; and eventually in 1917, the first Professor of Geography, on the creation of the John Rankin chair. Professor R. H. Kinvig, a pupil of Roxby's has said: 'Never did a university have a more devoted, sincere, or loyal servant, and never did a geography department have a more inspiring head. He never spared himself, and he built up what came to be acknowledged as one of the finest departments in the country (Plate 14 (*a*)).'[6]

Like Mackinder, Roxby spent much time, especially from 1908 onwards, in giving University Extension lectures in Merseyside. When World War I broke out he found the need for lectures of this kind to be even greater. Writing of that war, Roxby said: 'It is the first in which the British Democracy has been directly concerned, and not least among the far-reaching effects which it has already produced on our national life is that of interesting it profoundly in European politics and in the problems of European nationalities. Any undue absorption in economic questions at home to the exclusion of international issues, has gone, probably for ever. Not easily will the wider Vision be lost. The immense demand for information and lectures on European history in relation to the war is one aspect of the change.'[7]

Roxby as a young man was called on to teach a pioneer subject in a pioneer university of unusual character, and in a city that was equally singular. There were no ancient Grammar Schools as in Manchester, Leeds, and Birmingham. There was a long tradition of sectarian bitterness. The social cleavage between dockland and suburbia was deep and real. Many of the students were able but impecunious and grants were small. This was the setting in which Roxby, with his innate love of humanity, was able to make a great contribution through the teaching of geography. But it was not until 1917 when the Honours School and the chair were established that Roxby was able to exert his real power. In the end he became a sort of legend in the whole of Merseyside. By 1925 there can have been few teachers in Merseyside schools who had not heard of him. The Liverpool Honours School of Geography was the first in Britain but it was quickly followed in the same year by Aberystwyth under Professor H. J. Fleure, and by London in 1919. The Liverpool School was one of marked individual character but it derived some features from the Oxford Diploma. It included four principal elements. First there were general or systematic courses in both physical and human geography, the latter being an especially important feature. In the second place there was a complete course on the historical geography of the British Isles, not surprising in view of Roxby's training. Thirdly there were regional studies of areas such as China and the USA in which both physical and human aspects were studied, while as much as possible of the historical geography was included as an essential background to the present human geography. Lastly there was always a sound study of cartography.

Roxby's first geographical interest was in historical geography and especially of certain rural areas in Britain. How did his great knowledge of, and consuming passion for, China begin? In 1912 Roxby was awarded an Albert Kahn Travelling Fellowship. These were of the value of £660, a large sum in those days, and had been founded by M. Albert Kahn of Paris to enable the persons appointed to travel round the world. Two Fellows were to be appointed each year; Lord Curzon, President of the Royal Geographical Society in 1911–14, was one of the first trustees. The first two fellows were elected in 1910. Roxby and G. Lowes

Dickinson, Fellow of King's College, Cambridge, were appointed in 1912. It is interesting to notice that A. G. Ogilvie, another Oxford geographer, was elected in 1914, when Junior Demonstrator in Geography at Oxford, but never took up his fellowship because of the war.

In 1912 Roxby was Lecturer in Regional Geography in the University of Liverpool and he held the Kahn Travelling Fellowship from September 1912 to September 1913. It was part of his duty to write an account of his journey. His *Report to the Trustees*, a pamphlet of seventy-four pages, was published in 1914. In the *Introduction* (p 3) Roxby wrote: 'As my journey proceeded the conviction forced itself upon me more and more that the interrelation of East and West is to be one of the central issues of world politics in the twentieth century.' Roxby travelled through Italy, spent three weeks in Egypt, reached Bombay at the end of November and spent three months in India and Ceylon. He visited Rangoon and Lower Burma, spent a fortnight in Singapore, then went to Java, Hong Kong, Shanghai, Nanking, and Hankow. He stayed a month in Peking and went on to Korea and Japan. In his *Introduction* (p 4), Roxby says: 'Any real change in the economic structure of society involves sooner or later, not merely a corresponding change in social and political relations, but a revaluation of religious ideas . . . I believed that the spiritual orientation of a community was immensely the most important aspect of its life, especially in India, but even there it could hardly escape from the influence of a changing economic environment.' This exemplifies Roxby's deeply religious attitude to his teaching and research. His *Report* consists of three parts: (1) 'Some impressions of Modern Italy' (pp 7–14) which is followed by a section on 'the Church and Social Development in France and Italy' (pp 15–21); (2) 'India' (pp 22–54); and (3) 'China' (pp 55–74). The pages on China show how deeply Roxby was influenced by that part of his tour. It is not surprising that Roxby gave so much of the following thirty years of his life to teaching and writing about China. On p 56 he writes: 'China holds a unique place in the impressions which my memory stores of a crowded year of travel. No country, not even India . . . so captivated my imagination.'

He gives three reasons for the deep impression which China made upon him: 'The first was an instinctive liking for the people, a liking as great as that which I felt towards the Burmese.' The second was 'that China came more as a revelation to me than any other country . . . I had never been able to conjure up a real image of the celestial land. Of all countries, China seems to me the hardest to realize until one has actually seen it . . . no journey more altered my perspective of world politics as the two months which I spent in the Great Flowery Land . . . The third factor which contributed to create in my mind a special interest in China was one's constant feeling of the uncertainty of its future and of the terrific issues which are involved in its destiny'. The *Report* analyses the conflict between western and eastern cultures as seen in India and the Far East. This was the beginning of Roxby's life-long devotion and attachment to China and its people.

In 1921–2 Roxby paid another visit to China when serving on the China Education Committee. In 1944 he resigned his Liverpool chair and went to China as the chief Representative of the British Council, and died in that country on 17 February 1947 at the age of sixty-six. This decision to go to China in old age, so soon after his happy marriage in 1941 to Marjorie Peers Howden and his purchase of a house for retirement in his native Buckden, must have been motivated by his devotion to China and its people.

What did Roxby write on China? First he produced a number of articles some of which will be discussed later. Then in 1920 he wrote a short monograph, *The Far Eastern Question in its Geographical Setting*, published by the Geographical Association. Lastly, during the war he edited the volumes on China in the series of *Geographical Handbooks* published by the Admiralty. *China Proper*, in three volumes, appeared in 1944 and 1945; a large part of volume 1 was written by T. W. Freeman and others under Roxby's editorship. Roxby had long planned to produce a large-scale work on China, but he found that the rigid framework of the Admiralty's official scheme cramped his style.

Roxby's most important article on China was 'the distribution of population in China'.[8] He had read this to Section E of the

British Association at Toronto in August 1924. The article was based on a survey volume by the China Continuation Committee entitled *The Christian Occupation of China* (1922), of which only a few copies were published. Roxby's first article on China appeared in the *Scottish Geographical Magazine* and dealt with 'Wu-Han', the heart of China, the three cities of Hanchow, Wuchang and Hanyang.[9] In this article, published in 1916 during World War I, Roxby deplored the British Government's refusal to support Lord William Cecil's scheme to establish a university under British auspices in Central China at Hankow or Wuchang. The Government's refusal was based partly on the ground that a British university had already been established in Hong Kong, but the main reason was the indifference of the British public to the scheme. Before the war America, Japan and Germany had given money not only for educational work in China, but also for Chinese students to work in their own universities in the three countries named. 'In both these respects,' wrote Roxby, 'Great Britain has lagged a long way behind these countries. This indifference seems to be singularly short-sighted from the point of view of our own commercial interests since Chinese students who afterwards become merchants, manufacturers or engineers naturally tend to build up a trade connection with the country under whose auspices they have been trained. It is to be regretted too since . . . Great Britain is in some respects better qualified than any other country to help China in the solution of the critical educational problems with which she is now confronted.'

Roxby's contribution to the theory of regional geography must be considered briefly. He was, of course, a disciple of Herbertson and a firm believer in regional geography. In 1907 he followed up Herbertson's famous 1905 paper with one of his own 'What is a Natural Region?'[10] In this he boldly declared that '*ceteris paribus* a particular set of physical conditions will lead to a particular type of economic life. A physical unit tends to become an economic unit'. Again in 1926 Roxby persisted in using the term 'natural' for such an entity as Central Europe, despite the significance of cultural and other factors in its evolution on the ground that 'one ought to think of Man as part of Nature in Regional Geography.'[11]

In this paper he emphasised the importance of a factor not often taken into account. This he called 'the space relationship of regions, i.e. their external relations, particularly in human affairs, as distinct from their internal or intrinsic conditions'. This significant idea was a feature of his work on China.

But Roxby was a teacher rather than a writer. To him 'The primary object of geographical education was the development of citizenship in the fullest sense, both at home and in the world at large'.[12] Roxby regarded it as the duty of the regional geographer to comment on the possible future uses of land, as he did in China. Both Hakluyt and Mackinder had the same belief that geographers should concern themselves with the future: this tendency seems to be strongly ingrained in Oxford geographers.

If I were asked which geographical publication of Roxby's should be read by all students I would suggest his Presidential Address 'The scope and aims of human geography' which he delivered to Section E of the British Association at Bristol in 1930.[13] This is still in many respects up-to-date. It is also beautifully composed—Roxby took immense pains with his English. It was largely because he was a perfectionist that he wrote so little. I will quote the final paragraph of Roxby's Bristol lecture because it sums up so much of his teaching.

> We may claim for human geography that, rightly studied, it is a vital element in training for national and international citizenship. It can enable us 'accurately to imagine the conditions of the great world stage' and the place of the different regions within it. It is a valuable mental discipline, calling for an exact sense of proportion in appraising the value of many factors and more specifically developing the great quality of sympathetic understanding. The point of view and type of outlook which it fosters were never more needed than in the present critical stage of human development. Yet not only through its value as an educational instrument, but also through the programme of constructive work which it advocates, can it contribute to the realisation of the ideal of 'unity in diversity', and that seems the only possible ideal for the life of humanity on a planet, which, however small

applied science may make it, will always retain its infinite variety.

Roxby's teaching built up a school of geographical thought at Liverpool and one that was based ultimately on the work of his own Oxford teachers, Mackinder and Herbertson. Many of Roxby's former pupils became Professors and Readers in universities and spread his influence widely, especially in the North of England. The following geographers were once Roxby's pupils: R. H. Kinvig (Reader and later Professor at Birmingham, 1924–58); A. V. Williamson (Professor at Leeds, 1928–52); W. Fitzgerald (Professor at Manchester, 1944–8); Wilfred Smith (Professor at Liverpool, 1950–5); H. King (Professor at Hull, 1946–58); H. R. Wilkinson (Professor at Hull, 1958–present); S. J. K. Baker (Professor at Makerere from 1946); and George Tatham (Professor of Geography at York University, Toronto). The influence of Roxby's thought, especially of his views of regional geography, has been far-reaching. His influence on historical geography has also been great, but he has received little credit for his teaching in this branch of geographical learning. Yet he always viewed geography from a historical standpoint and often quoted Mackinder's remark that 'the present is the past flowing into the future'.

Roxby's love for individuals was instinctive. At Liverpool he took a deep personal interest in foreign students, especially in each Chinese and Egyptian.[14] Roxby regarded the pastoral side of his work as University teacher as being of primary importance in his life. He was the active helper of all Christian causes in the University, the City and the Diocese of Liverpool. Roxby liked people; this was the key to his relationship with individuals but also to his study and teaching of geography.

Professor J. F. Unstead once remarked that Roxby's School of Geography 'was almost like a family; there were unusually close ties between himself, his staff and his students—a relationship which was very helpful on both the personal and the professional sides. I vividly remember how, when I was his external examiner at Liverpool, and we came together to the department to conduct the "orals", a group of students waited at the door and Roxby, noticing an expression of nervousness, put his hands on the shoulders of one of them and said: "Come along, old man, let's get it over".' Unstead continues: 'the same feeling of friendship made his first visit to China . . . the beginning of a close association between himself and the Chinese people which lasted till his death among them in 1947. He thought of China, not merely in geographical terms, but as the home of real people; he studied their history and their modes of life, sympathised with their suffering under natural and man-made calamities, admired their culture. On all my visits to Liverpool I met Chinese students who had come to learn from him and who, in their turn, effected a valuable widening of the mental horizons of his department'. Professor H. J. Fleure spoke in very similar terms of the Liverpool School of Geography: 'the family life of Roxby's department at Liverpool was one of its greatest features and he was not so much the father as the elder brother with the emphasis on brother rather than on elder.'

Roxby was a superb lecturer. He was primarily a teacher and he stirred and stimulated all those who heard him. In the late Professor Wilfred Smith's inaugural lecture given at Liverpool on 1 November 1951 he wrote of Roxby: 'He ranged over all space and all time and, so it seemed to the undergraduate browsing in new pastures, all knowledge. He was a master in moulding a map to his use, in giving it life and speech, alternately caressing it and bludgeoning it till its meaning stood out bold and clear.' Roxby's impact as a lecturer and speaker in Merseyside in both extramural classes and schools was tremendous; he was extremely well known in the area and is still remembered today.[15] On 26 February 1965 Mr Harold Wilson, then Prime Minister, received the honorary degree of LLD of the University of Liverpool and opened a new Union building. The Public Orator's speech was printed in the Liverpool *Recorder*, but not the Prime Minister's reply. However, a former Oxford geographer, who was present on that occasion, has told me that the Prime Minister, in his reply, remarked that when he was at Wirral County Grammar School (still a flourishing school in the Borough of Bebington), his only contact with the University of Liverpool had been to hear Roxby give a lecture on

China—a lecture so forcibly delivered that the outline remained with him still.

Roxby was very tall, about 6ft 4½in. He was clean-shaven and had a pointed chin with a strong jaw and a broad but relatively low brow and grey eyes. He stooped slightly and was sometimes a little jerky in movement; his general appearance was rugged. He was cheerful and genial and had a certain carelessness in dress. Almost invariably he wore a grey tweed suit. He had a commanding presence but his manner was easy and unaffected and he was invariably courteous. The pattern and shape of his lectures could be discerned easily; normally his lectures were given without any notes. It was obvious to any listener that his lectures had been hammered into shape during conscientious and thorough preparation. He had the ability to establish friendly contact with his audience from the start and it was immediately apparent that he was a complete master of his subject. He always carried his hearers with him—and his audiences were often very large. He gave the impression, without being pompous or acting a part, that what he said was supremely worth saying and knowing. His sincerity was obvious and effortless—part of the man himself.[16]

I can myself confirm J. E. Allison's opinions, given in the previous paragraph, as I heard Roxby deliver two addresses both of which made a deep impact on me. In 1924 and 1925 I was teachin the University of London and was far from happy in my work as a geographer. I had lost faith in geography and had grave doubts about its value. I was seriously considering a change from geography to the law and had taken some examinations with this end in view. One Saturday morning I went to hear Roxby speak on China to a branch meeting of the Geographical Association in central London. I was so moved by that speech, given with such sincerity when he was at the height of his powers, that I abandoned all ideas of the law as a profession. I think Roxby's obvious certainty that his work was a part of his religion was the cause of the tremendous impact he made on his hearers. One could sense his spiritual outlook; and his passionate desire that geography should be a force for world peace. I heard him again in 1930 when he delivered his famous address on 'the scope and aims of human geography' which only served to increase my admiration of his ability and character.

About two thousand years ago the geographer Strabo explained that geography was the concern of the philosopher in these words:

No one can undertake a work on geography without the wide learning only possessed by the student of things human and divine, the knowledge of which men call philosophy. And again, the usefulness of geography—and it has varied uses, not only for politics and war, but also in giving knowledge both of the heavens and of things on land and sea, animals, plants, fruits, and of all that is to be seen in different regions —likewise presupposes the philosophic mind of one who studies the art of life, that is, of happiness.'[17]

Hakluyt, Mackinder, Herbertson, and Roxby each had his own individual philosophy, but they were all students of the whole environment in which men have their being. They all studied 'the art of life, that is, of happiness' and they found happiness in their life's work. Each of them found the study of the earth as the theatre of human life to be an absorbing, unending but happy task.

NOTES

1 I am indebted to the late Canon Charles F. H. Soulby of Liverpool Cathedral, who in 1951 provided me with a short MS biography of Roxby. An edited copy of this MS was deposited in the School of Geography, University of Liverpool. I am also grateful to J. E. Allison, a pupil of Roxby's in the period 1911–14, and later chairman of the Liverpool and District branch of the Geographical Association, who gave me much detailed information about Roxby and his life in Liverpool. The obituary notices of Roxby are slight; in my opinion Roxby's services have not been adequately acknowledged outside the North of England. In 1967, when the fiftieth anniversary of the foundation of the Liverpool Honours School of Geography, the first in Britain, was celebrated,

Professor R. W. Steel wrote a chapter (1–23) on 'Geography at the University of Liverpool' in *Liverpool Essays in Geography* (1967). In this essay Professor Steel gives a brief account of Roxby's work. Roxby was essentially an Oxford man. He was deeply influenced by Mackinder and Herbertson for both of whom he had an affectionate regard; it was his version of their gospel that he disseminated in the North of England. This chapter was delivered as last of four valedictory lectures at Oxford, on 17 May 1967. It was an outside contribution to the jubilee celebrations at Liverpool. Oxford should be proud that the foundations of the great School of Geography on Merseyside were securely laid by one of its sons. T. W. Freeman, who at one time worked closely with Roxby wrote about him in chapter 7 (156–68) of *The Geographer's Craft* (Manchester University Press, 1967).

2 The following papers are examples of Roxby's interest in the historical geography of rural England. 'Historical geography of East Anglia', *Geog Teacher*, 4 (1908), 284–92; 5 (1909), 128–44; 'Rural depopulation in England during the nineteenth century', *Geog Review*, (1912), 174–90; 'the agricultural geography of England on a regional basis', *Georg Teacher*, 7 (1914), 316–21.

3 Smith, Wilfred, *Geography and the Location of Industry* (1952), 2.

4 Ogilvie, A. G. (ed.) *Great Britain* (1928), chap 8. 'East Anglia', 143–66.

5 Professor John Frederick Unstead (1876–1965). Obituary by E. W. Gilbert in *Geogr J*, 132 (1966), 334–5.

6 Kinvig, R. H., 'The Geographer as Humanist', Presidential address to section E of the British Association at Liverpool, 1953. This address includes much useful information about Roxby.

7 Roxby, P. M., *Report to the Albert Kahn Trustees* (1914), 7.

8 Roxby, P. M., *Geog Rev*, 15 (1925), 1–24.

9 'Wu-han: the heart of China', *Scott Geogr Mag* 32 (1916), 266–78. Other important contributions by Roxby to the geography of China were 'the terrain of early Chinese civilization; *Geography* 23 (1928), 225–36; 'China as an entity: the comparison with Europe', *Geography*, 19 (1934), 1–20; and 'the expansion of China', *Scott Geog Mag*, 46 (1930), 65–80. Roxby also wrote a pamphlet on *China* in the series Oxford Pamphlets on World Affairs no 54 (1942).

10 'What is a natural region?', *Geog Teacher*, 4 (1907), 123–8.

11 'The theory of natural regions', *Geog Teacher*, 13 (1926), 376–82.

12 Roxby developed this theme in 'Geography in citizenship' in *Geography in Education* (1919), 37–40.

13 *Scott Geog Mag*, 46 (1930), 276–90.

14 Roxby spent one term of 1930 as Visiting Professor at the Egyptian University of Cairo and thus established links with Egypt.

15 Roxby sometimes wrote on the geography of Merseyside; an especially useful paper was 'Aspects of the development of Merseyside', *Geography*, 22 (1927), 91–101.

16 This paragraph is largely based on accounts written by J. E. Allison, a pupil of Roxby in 1911–14.

17 *The Geography of Strabo*, i, I, I. This translation was made for me by the late Sir Richard Livingstone, then President of Corpus Christi College, Oxford. It is printed at the conclusion of my inaugural lecture at Oxford in 1954 (*Geography as a Humane Study*, Oxford, 1955).

CHAPTER TWELVE

Vaughan Cornish (1862–1948) and the Beauty of Scenery

O^N 26 FEBRUARY 1957, the University of Oxford accepted
by Decree a bequest of the residue of the estate of Dr
Vaughan Cornish and accorded its appreciation of the
generosity of the testator. A further Decree was passed by Con-
gregation on 29 April 1958 by which regulations were made for
the administration of the bequest. In accordance with the terms of
the will the income of 'an endowment to be known as "The
Vaughan Cornish Bequest" . . . shall be devoted to the encouragement and assistance of postgraduate students of the University
engaged in the advancement of knowledge relating to the beauty
of scenery as determined by nature or the arts in town or country
at home or abroad'. A Board of Management was set up and during the past twelve years a number of grants have been made to
members of the University engaged in studies covered by the
terms of the bequest. A high proportion of the grants have been
awarded to geographers.[1]

Dr Vaughan Cornish was not an Oxford graduate and he
appears to have had little contact with either the University or
the City of Oxford. Yet he entrusted this endowment to the
University for a cause that was very dear to him. *The Times* in a
long obituary, headed with the words 'Geography and Natural
Scenery' described Cornish as 'a geographer who gained a special
place in the scientific world for his study of land and water waves
and who devoted much of his energy in later years to the appreciation and preservation of natural scenery'. In *Who's Who* Vaughan

Cornish entered his occupation as 'Geographer' and this is a true description. In the University of Oxford the name of Vaughan Cornish, a graduate of the University of Manchester, is not widely known, but it seems right, not long after the centenary of his birth, to commemorate a benefactor and to look for the reasons which prompted him to make such an endowment.[2]

I will divide this chapter into two parts. First I will give a brief account of Vaughan Cornish's life and work. I will then outline the present position with regard to the conservation of the beauty of scenery in England and Wales. I want to emphasise the great need for more research of this subject, and thus indicate the value of the Vaughan Cornish Bequest.

Vaughan Cornish was born on 22 December 1862, at Debenham in Suffolk (Plate 16). He was the third son of a clergyman, then Vicar of Debenham, but the Cornish family had been connected with the Sidmouth district of Devonshire for several centuries. Vaughan Cornish later acquired a strong attachment for that part of England and maintained his contacts with it until the end of his life. He went rather late to St Paul's School where he became interested in the natural sciences. He then proceeded to Owens College, Manchester, partly because it offered facilities for research in chemistry. He took his BA at Manchester University in 1888, and proceeded to the degree of DSc in 1901.

For a number of years Cornish was Director of technical education to the Hampshire County Council, but in 1895 he resigned this post and devoted himself to his own researches. In 1891 he married Ellen Agnes Priors, daughter of Alfred Priors, an artist of Kingston Lisle, Berkshire. Private means enabled the couple to travel extensively; indeed they gave up a settled home in order that he might devote himself entirely to research. This work consisted mainly of studies of physical phenomena, such as surface waves of all kinds, the Severn Bore and the Trent Eagre, the length at speed of waves observed on voyages across the Atlantic, the action of wind on snow, and the formation of dunes in the desert. He produced a number of papers on subjects of this kind and in 1900 the Royal Geographical Society awarded him the Gill Memorial Prize. Later he published his *Waves of Sand and Snow*

(1913) and *Ocean Waves and Kindred Geophysical Phenomena* (1934). Vaughan Cornish and his wife were in Kingston, Jamaica, at the time of the earthquake of 1907; they were lucky to escape with their lives. He contributed two papers on the Jamaica earthquake to the *Geographical Journal*.[3]

Vaughan Cornish developed an interest in another branch of geography by writing between 1910 and 1914, three papers on the Panama Canal. He continued to work in historical and strategical geography; and during the war of 1914–18 he lectured to officers in the services on such subjects as lines of communication, national resources and maps. Partly as a result of these studies he published in 1923, his largest book, *The Great Capitals*, which has been described as 'the strategical geography of the past'. In his discussion of political capitals Vaughan Cornish introduced his idea of 'natural Storehouses', 'Crossways' and 'Strongholds' as categories.[4]

However, the most significant part of Cornish's work concerns the beauty of scenery. He was undoubtedly inspired to devote the main energies of the later part of his life to this type of work as the result of an address given by Sir Francis Younghusband (1863–1942). In order to understand the impact of this speech on Vaughan Cornish, it is necessary to give an account of the occasion and to quote at some length from Younghusband's remarks. On 31 May 1920, Sir Francis Younghusband, then President of the Royal Geographical Society, addressed the Society's anniversary meeting on the subject of 'Natural Beauty and Geographical Science'.[5] Sir Francis was an intrepid traveller and a great authority on Central Asia. He is remembered for his famous march to Lhasa in 1904 to negotiate a treaty with Tibet.[6] His speech is still worth reading in full: I will quote a few passages from it in order to give some idea of his main theme. 'While Geology', said Sir Francis, 'concerns itself with its (the Earth's) anatomy, Geography, by long convention restricts its concern to the Earth's outward aspect. Accordingly, it is in the face and features of Mother-Earth that we geographers are mainly interested. We must know something of the general principles of geology, as painters have to know something of the anatomy of

the human or animal body. But our special business as geographers is with the outward expression. And my second point is
that the characteristic of the face and features of the Earth most
worth learning about, knowing and understanding is their beauty;
and that knowledge of their beauty may be legitimately included
within the scope of geographical science.

'It may be argued, indeed, that science is concerned with
quantity—with what can be measured—and that natural beauty is
quality which is something that eludes measurement. But Geographical science, at least should refuse to be confined within any
such arbitrary limits and should take cognizance of quality as well
as quantity.'[7] Again, later in his lecture he said: 'Man is both
affected by the beauty of Earth's features and himself affects that
beauty. And this relationship between man and the natural beauty
of the Earth is one which Geography should take as much cognizance as it does of the relationship between man and the productivity of the earth.'[8]

Towards the end of his lecture Sir Francis explained how these
ideas could influence the work of the Royal Geographical Society.
'We are constituted as a Society', he said, 'for the purpose of
diffusing geographical knowledge, and I trust that in future we
shall regard knowledge of the beauty of the Earth as the most
important form of geographical knowledge that we can diffuse'.[9]
He elaborated this argument in the following passage: 'We shall
think travellers barbaric if they continue to concern themselves
with all else about the face of the Earth except its beauty. We shall
no longer tolerate a geographer who will learn everything about
the utility of a region for military, political, and commercial purposes, but who will take no trouble to see the beauty it contains.
We shall expect a much higher standard of him. We shall expect
him to cultivate the power of the eye till he has a true eye for
country—a seeing eye; an eye that can see into the very heart and
through all the thronging details, single out the one essential
quality; an eye which cannot only observe but can make discoveries. We shall require him to have the capacity for discrimi-
nating the essential from the unessential, for bringing that essential
into proper relief and placing upon it the due emphasis. When he

thus has true vision and can really see a country and when he has acquired the capacity for expressing either in words or in painting what he has seen, so that he can communicate it to us, then he will have reached the standard which this Society should demand. And this is nothing less than we expect of him that he should have in him something of the poet and the painter'.[10]

Vaughan Cornish was fifty-seven when he heard this address and, largely in response to Sir Francis Younghusband's appeal, now turned his main attention to the 'analytical study of beauty in scenery'. He seems to have been particularly impressed by one sentence of Younghusband's, which he himself used in 1928 as a text to his own Presidential Address to the Geographical Association. 'What men naturally do, and what I would suggest Geography should deliberately do, is to compare the beauty of one region with the beauty of another so that we may realize the beauty of each with a greater intensity and clearness'.[11] For the remaining years of his life Vaughan Cornish devoted a large part of his energies to the study of the beauty of scenery; but he also made practical efforts to conserve the beauty of existing scenery. Much of his work he regarded as educational; in the preface to *The Beauties of Scenery. A Geographical Survey* (1943) he argued that wartime was an appropriate time for his manual to appear, 'on account of the need for education in scenic amenity in preparation for the re-planning of town and country, in this and other lands, when peace has been restored'.[12]

It is impossible to give a detailed account of all Vaughan Cornish's writings, but a list of the titles of some of his books will give an idea of their content. *National Parks* appeared in 1930 and was followed by *The Poetic Impression of Natural Scenery* (1931), *The Scenery of England* (1932), *Scenery and the Sense of Sight* (1935), *The Preservation of our Scenery* (1937), *The Scenery of Sidmouth* (1940), and *The Beauties of Scenery: A Geographical Survey* (1943). These books appeared alongside a stream of papers. In 1925 he read a paper to the British Association at Southampton on 'Apparent magnitude in natural scenery and its determining causes'; and in 1926 he addressed the Geographical Association on 'Rhythmic motion in rivers: a study in scenery'. Two of his presidential addresses had similar themes, 'Harmonies of scenery: an outline of aesthetic geography', delivered to the Geographical Association in 1928, and 'The Scenic Amenity of south-eastern England' given to the South-Eastern Union of Scientific Societies in 1935. He also campaigned actively for the Council for the Preservation of Rural England; he was a member almost from its beginning in 1926. He was early interested in the movement for National Parks and gave evidence on behalf of CPRE, to the National Parks Committee in 1929. In 1936 he addressed the British Association at Blackpool on 'National parks and the preservation of nature in England'. He became especially interested in the preservation of the coast. In 1935 he appealed to the South-Eastern Union of Scientific Societies to help to secure the cliff edge against building and against enclosure by garden fences. 'I trust', he said, 'that members . . . will use their influence to secure an open, unenclosed pedestrian broadway a hundred yards in width along all the lofty cliffs of Southern England. Let it never be forgotten that the sea view from the cliffs is the special scenic heritage of our island people.[13] In his book on *The Scenery of Sidmouth: Its Natural Beauty and Historic Interest* (1940) there is further evidence of his opinions on this subject. In the preface to this book he wrote: 'Since I first knew the Sidmouth district in the days of childhood I have been a pilgrim of scenery in many lands, but no aspect of the world from Arctic to Equator has diminished my admiration for the beauty of this peaceful spot on the south coast of Devon'. Vaughan Cornish died at Camberley on 1 May 1948. In the following year the National Parks and Access to the Countryside Act came into force. Vaughan Cornish did not live to see the apparent victory of the cause for which he had battled so long. G. R. Crone has rightly said that Cornish had 'a somewhat austere bearing, aloof and reserved but courteous'. (*DNB*) I remember that, although Cornish was shy, he was very ready to encourage and help much younger men.

The forty years after Younghusband's famous address (1920) saw the rapid rise of the motor car as a means of transport; in 1922 there were less than one million motor vehicles of all kinds in Great Britain. This revolution in means of communication has involved an every-increasing pressure on the countryside, and was responsible, in the 'thirties, for building in ribbons along roads outside towns and villages; as well as for much development of the coast. The need for the protection of the scenery of Britain by official action during the years between the wars became progressively greater, but very little was done. In 1931 the official Report of the National Park Committee, under Christopher (later Lord) Addison as chairman, pronounced in favour of the establishment of 'National Reserves', areas of exceptional natural interest which were to be safeguarded 'against disorderly development and spoilation'. In spite of this Report nothing was attempted; the 'thirties were not a propitious time for defending amenity.

Since 1945 a variety of practical measures have been taken to protect defined areas of land against harmful development; they can be grouped under five heads. (Fig 19 (*a*)).

1 NATIONAL PARKS. The first result of the National Parks and Access to Countryside Act of 1949 was the setting up of a National Parks Commission. Since 1951 ten National Parks have been delimited and established by the Commission (Fig 19). In all they cover 5,254 square miles, that is roughly 9 per cent of England and Wales; more than a quarter of a million people still live in the Parks, but in the main they can be regarded as areas of rural depopulation. The Commission has two main duties; first 'the preservation and enhancement of natural beauty'; and second the provision 'of opportunities for open air recreation and the study of nature by those resorting to National Parks'. As part of its first duty the Commission can make representations to Ministers and local planning authorities on development 'likely to be prejudicial to the natural beauty of any area,' whether in a Park or not.[14] By the Countryside Act of 1968 the National Parks Commission was replaced by a Countryside Commission with wider powers.

2 AREAS OF OUTSTANDING NATURAL BEAUTY. These form a lower tier of National Parks; they are smaller in extent and are usually without wide stretches of open country, and consist in the main of agricultural land (Fig 19 (*a*)). In 1956 the first of these areas (Gower) was established by the Commission. In all seventeen Areas had been designated and confirmed before September 1964;

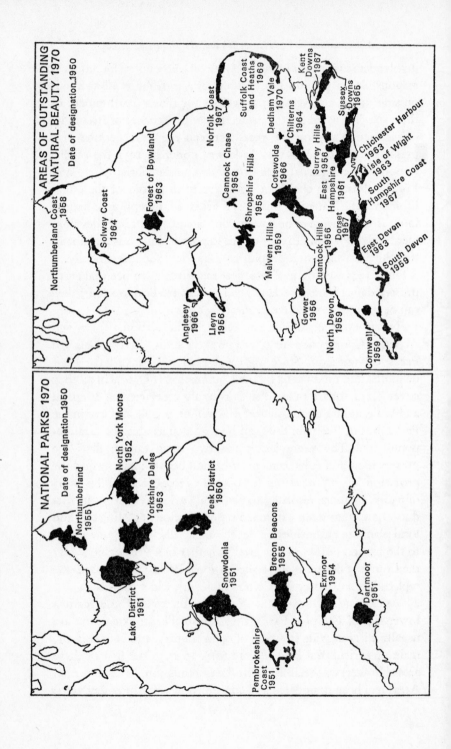

NATIONAL PARKS 1970

Date of designation..1950

Northumberland 1955

North York Moors 1952

Yorkshire Dales 1953

Peak District 1950

Lake District 1951

Snowdonia 1951

Brecon Beacons 1955

Pembrokeshire Coast 1951

Exmoor 1954

Dartmoor 1951

AREAS OF OUTSTANDING NATURAL BEAUTY 1970

Date of designation..1950

Northumberland Coast 1958

Solway Coast 1964

Forest of Bowland 1963

Norfolk Coast 1967

Cannock Chase 1958

Suffolk Coast and Heaths 1969

Shropshire Hills 1958

Cotswolds 1966

Dedham Vale 1970

Chilterns 1964

Kent Downs 1967

Malvern Hills 1959

Surrey Hills 1956

Sussex Downs 1965

East Hampshire 1961

Quantock Hills 1956

Dorset 1957

Chichester Harbour 1963

Isle of Wight 1963

South Hampshire Coast 1967

East Devon 1963

South Devon 1959

Anglesey 1966

Lleyn 1956

Gower 1956

North Devon. 1969

Cornwall 1959

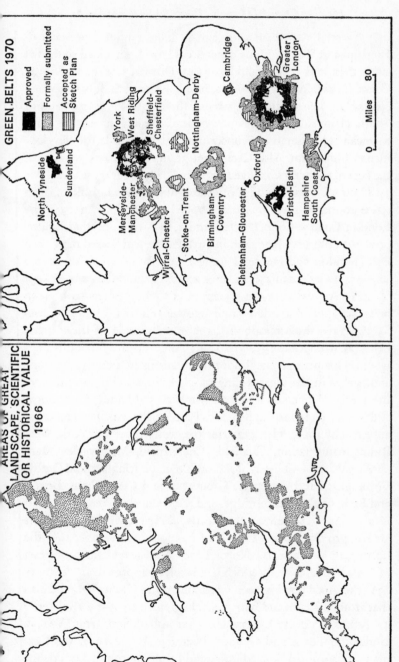

Fig 19 (a) and (b) Conservation Land (based on a map by the Department of the Environment used with the permission of the Controller of Her Majesty's Stationery Office). National Parks, Areas of Outstanding Beauty and Green Belts are marked as in 1970. No data later than 1966 are available for Areas of Great Landscape, Scientific or Historical value

GREEN.BELTS 1970

■ Approved
▨ Formally submitted
▥ Accepted as Sketch Plan

North Tyneside
Sunderland
York
West Riding
Sheffield-Chesterfield
Merseyside-Manchester
Wirral-Chester
Stoke-on-Trent
Nottingham-Derby
Cambridge
Birmingham-Coventry
Cheltenham-Gloucester
Oxford
Greater London
Bristol-Bath
Hampshire South Coast

0   80
MILES

AREAS OF GREAT
LANDSCAPE, SCIENTIFIC
OR HISTORICAL VALUE
1966

they covered a total of 2,367 square miles (about 4 per cent of England and Wales). Many other areas have been added to the list since then and these include Sussex Downs, Chilterns, Solway Coast, Cotswolds, Kent Downs, Norfolk Coast, Suffolk Coast and Heaths, Dedham Vale and parts of the coast of Anglesey and of South Hampshire. The Countryside Commission has the right to make representations concerning any proposals for development within them. Also Exchequer grants of up to 75 per cent of the cost of effecting improvements can be made for these areas.[15]

3   GREEN BELTS. The idea of establishing a green girdle of open space round a town was first applied to London. In 1935 the London County Council acted under its own somewhat limited powers to preserve land; and in 1938 Parliament passed the Green Belt (London and Home Counties) Act. By means of these two schemes no less than 35,500 acres around London have been kept open. In August 1955 the system of encircling towns with green belts was applied to other English towns (Fig 19 (*b*)). Mr Duncan Sandys gave three reasons for the establishment of these green belts, (*a*) 'to check the growth of large built-up areas' like Birmingham; (*b*) 'to prevent neighbouring towns from merging into one another' as in the West Riding of Yorkshire; and (*c*) 'to preserve the special characteristics of a town such as Oxford, Cambridge, and York'. The fourteen provincial green belts which had received approval by 1965, at least in principle, were: Tyneside, York, West Riding conurbation, Sheffield—Chesterfield, Merseyside—Manchester, Wirral—Chester, Stoke-on-Trent, Nottingham and Derby, Birmingham and Coventry, Gloucester and Cheltenham, Bristol and Bath, Oxford, Cambridge, and the Hampshire South Coast. It is important to remember that nearly all the provincial green belts are still provisional and await confirmation. In December 1964 the only exceptions were the North Tyneside area and a small area east of Manchester, both of which had received approval. In February 1965 the green belt between Cheltenham and Gloucester, and also that around Bristol and Bath were also approved by the Minister.[16]

4   AREAS OF GREAT LANDSCAPE, HISTORIC OR SCIENTIFIC VALUE. Under the Town and Country Planning Act of 1947, the local planning authority (the counties and county boroughs) can define

and propose such areas in their development plans for the Minister's approval (Fig 19 (*b*)). The amount of legal protection from development which these areas can receive is very flimsy in practice. These areas attract no grants; the National Parks Commission does not have to be consulted about them, but the local planning authorities, when considering applications for development within them, should give special consideration to the probable effect on the landscape. Nevertheless, they cover in all a large part of the map of England and Wales. A selection of them are likely to receive, in due course, the legal status of 'Areas of outstanding natural beauty'. Fig 19 (*b*) shows the position in 1966.

These areas are shown on the maps of Development Plans prepared by the counties and county boroughs by an open hatch and the appropriate initial letters referring to Landscape, Historic or Scientific Value. Circular 92, issued in January 1951 by the Ministry of Housing and Local Government includes the following relevant sentences concerning the definition of these areas:

> With regard to the general notation, some of the areas of great landscape and other value which planning authorities will wish to show by this means may at a later stage become the subject, wholly or in part, of proceedings under the Act of 1949. [ie National Parks, Nature Reserves, Areas of Special Scientific Interest, Areas of Outstanding Natural Beauty.] The marking of the development plan in such cases will in no way prejudice the consideration required under that Act.

> The Minister is aware that some of the areas which planning authorities are considering showing as of great landscape value are substantial in size. It seems evident that as landscape value is so much a matter of opinion and as a marking on the plan is bound to carry a restrictive connotation in relation to development the public are likely to take a highly critical interest in the size of the area and in the nature of possible restrictions. Generally, therefore, planning authorities (*a*) would be well advised to include as areas of great landscape value only those areas which seem likely to command general acceptance among the interested public; and (*b*) should indicate in the written statement their intentions with regard to

development control as plainly as the circumstances permit. The broad intention will normally be that a major consideration will be the effect of any proposed development on the landscape, but it is highly desirable that this should be expanded wherever possible to indicate how far the policy is likely to result in refusal of permission for specific sorts of development or in insistence on special attention to design and external appearance.

Where an area of landscape value lies only partly within the area of a planning authority it is hoped that special pains will be taken to work out a policy with the other local planning authorities concerned to which they can all adhere. From the point of view of the public it is obviously important that sharp differences should not arise in the treatment of an area merely because it falls within the jurisdiction of more than one authority.

5 THE NATURE CONSERVANCY. This organisation was set up by Royal Charter in 1949, and is directly responsible to the Lord President of the Council. The Conservancy has the power to acquire land for Nature Reserves which then receives Crown Land status.[17] Up to September 1963, 105 National Nature Reserves had received Crown Land status; these cover 218,000 acres. Local authorities have created nine nature reserves; the Forestry Commission and other public bodies have placed at the disposal of the Conservancy twelve Forest Nature Reserves. There are a number of private reserves and it is believed that a total of about 140 nature reserves exist in Great Britain; three-quarters of them are under the Conservancy. The work of the Conservancy is a scientific activity; it is not directly concerned with the beauty of scenery. Its functions are defined by Royal Charter in these words: 'to provide scientific advice on the conservation and control of the natural flora and fauna of Great Britain; to establish, maintain and manage nature reserves in Great Britain, including the maintenance of physical features of scientific interest; and to organise and develop the research and scientific services related thereto'. The Charter thus fulfilled some of the ideas advocated by Vaughan Cornish in his British Association paper in 1936 at Blackpool.[18] As

an example of the scientific nature of the Conservancy's work it can be noticed that its largest research station at Merlewood in the Lake District is studying 'the biological and physical processes in which the soils and the living organisms of a woodland are involved, and comparing these with moorland which was formerly under tree cover'.[19]

In addition 'Sites of Special Scientific Interest' are defined by the Nature Conservancy. Under Section 23 of the National Parks and Access to the Countryside Act, 1949, 'where the Nature Conservancy are of the opinion that any area of land, not being land for the time being managed as a nature reserve, is of special interest by reason of its flora, fauna, or geological or physiographic features, it shall be the duty of the Conservancy to notify that fact to the local planning authority in whose area the land is situated'. The degree of protection which these sites enjoy is less than the absolute protection given to the National Nature Reserves. In September 1962 there were 1,256 such sites in England and 242 in Wales.[20] Local examples near Oxford are Otmoor; Littleworth Brick Pits; Stanton Great Wood.

As has been said already the Nature Conservancy's work is purely scientific. This statutory body does not make comments on the impact of the modern industrial revolution on the countryside and its amenities unless it feels that its scientific interests are directly affected. However, the Conservancy sometimes cannot hide its general views on this subject. The following eloquent passage is taken from the introduction to an official report:

> Our country, as we used to know it, is vanishing before our eyes. We see its coasts and fields chosen as sites for houses and power stations. Its green hills sprout steel masts and pylons. Its old oakwoods are clear-felled. Its grasslands, once bright with wild flowers, are now ploughed up and are re-seeded. Its waters and its air are polluted. We seek peace and no longer find it. We know that progress has its price, but we all must wish that somehow ways may be found of enjoying it without losing so much of our national heritage. Our country's landscape arises from nature's responses to the land uses and land management of our forefathers. They lived with nature and

shaped the soil and water and the plant and animal life to their ways. Can we repeat their happy example in today's terms ?'[21]

All the above five categories of protected lands are defined by the State or by bodies set up by the State. Quite different from these lands are those protected by private organisations.

6 LANDS BELONGING TO CHARITABLE BODIES SUCH AS THE NATIONAL TRUST OR THE OXFORD PRESERVATION TRUST. The National Trust owns about 300,000 acres of land permanently for preservation as well as 150 historic buildings. The National Trust owns 165 miles of coastland and appealed, in 1965, to save about 900 miles of coast that was still unspoilt and considered worthy of preservation. The appeal was known as 'Enterprise Neptune'. Similarly the Oxford Preservation Trust's properties cover an area of 300 acres and include those on Boars Hill which were purchased to save the beauty of the landscape.

A total of about 23 per cent of England and Wales is protected if the two categories of National Parks are added to the Green Belts. It sometimes happens that the same land falls into two of the protected categories. But, if the Nature Reserves and the 'areas of great landscape, historic or scientific value' are added to the other three categories, it is believed that very nearly 40 per cent of the land in England and Wales enjoys protection of some sort from incongruous or harmful development. But this so-called protection is more apparent than real; it is certainly precarious in the 'areas of great landscape, historic or scientific value'.

Before I discuss the various kinds of 'intrusion' that have taken place into the protected lands I must briefly explain why it has become so difficult to give absolute protection to any area. This is a small island and its population is increasing fast, much faster than was anticipated in calculations made at the end of the war in 1945. Between 1931 and 1951 the net increase of population of England and Wales was 4·6 millions; between 1951 and 1961 it was 2·3 millions. In the middle of 1962 the Registrar-General estimated that the population of England and Wales was 46·7 million and he calculated that in the year 2002, it would be 63·7 million. This small and densely peopled island will become ever more crowded. The pressure not only on good agricultural land, but also on all the land now protected will become continually greater. In the ten years 1971–81 it is now expected that there will be an increase of population of well over 2·4 millions. Where are they to live? That is one of the main problems that face this country. It was officially forecast in 1961 that there will be a growth of at least 3½ million by 1981 in that part of England which lies south-east of a line from the Wash to Lyme Regis.[22] This increasing population is also increasing its mobility. In 1939 there were 3·1 million motor vehicles on the roads of Great Britain; by 1962 the number had increased nearly fourfold. Increased mobility inevitably means an increase in the pressure on the more isolated parts of the country. While the amount of land remains the same, more and more people are searching for space on which to live and work and play. Sir Christopher Hinton put the matter quite clearly: 'The United Kingdom is supporting a population of 50 million people in an island well designed to accommodate 20 million, and is seeking to achieve a continuous improvement in the material standards of living. This cannot be done without growing industrialisation, and growing industrialisation demands that areas which hitherto have been rural in their character shall be used for industrial purposes.'[23]

I will now outline some of the so-called 'intrusions' into the protected lands which have already taken place. The National Parks have not remained inviolate; this is not entirely surprising. Britain finds it far more difficult than does the United States, with its vast reserves of land, to prevent economic development in National Parks. It was not the original intention in 1948 that the British National Parks in all circumstances should remain inviolate, but it was certainly not expected that large-scale economic developments would occur. At Milford Haven, astride the eastern boundary of the Pembrokeshire Coast National Park two oil refineries and an oil terminal have been established in spite of the opposition of the National Parks Commission. In November 1957 the Minister of Housing and Local Government ruled that the Esso Petroleum Company might erect an oil refinery, the greater part of which was to be located in the National Park. He did not think that 'the damage to the amenity of the Park by the industrialisation of this small section' was 'sufficiently serious to warrant refusal of consent'. [24] In June 1958 the Minister gave permission for the establishment of a large iron-ore stocking ground at Angle, south of Milford Haven, but in the Park. The Parks Commission considered this project 'to be totally at variance with the right use of land in a National Park', [25] but the Minister did not consider that 'on the balance of advantages in the national interest he would be justified in refusing to grant the planning permission which the Company have sought'. [26] The Minister had to make a decision between two competing claims on land. On the one hand were the claims of amenity put forward by the National Parks Commission; on the other hand were the claims of the industry that the natural facilities of the great deep-water inlet of Milford Haven for the purpose of oil refineries were unique and had to be used. [27]

At Milford Haven it was a case of industrial development by large companies. In Snowdonia the 'intruder' was the Central Electricity Board. In July 1958 the Minister of Power announced that, after consultation with the Minister of Housing and Local Government he had decided to approve the Electricity Board's application to build a nuclear power station at Trawsfynydd Lake in the Snowdonia National Park and to erect the associated overhead lines. The station was expected to cost £50 million. The proposal had been opposed by the National Parks Commission on the ground that there must be 'the strongest presumption against the construction of large scale industrial installations in a National Park'. In its representations the Commission 'submitted that Trawsfynydd was the central point of that splendid panorama of mountains around Ffestiniog from Moelwyn Bach on the west to Manod Mawr on the east and that the steep face of this range, particularly from the south side of the reservoir, was one of the finest mountain prospects in North Wales. A nuclear power station in such a setting could not appear other than a huge and overmastering imposition, a creature utterly out of scale and character with the fine landscape around it. The introduction of anything as large as a nuclear power station at once upset the whole scale and emphasis of the landscape'. [28] The Report of the

National Parks Commission gives the following account of the Public Inquiry: 'In the Trawsfynydd case the fundamental question to be decided was, as the Inspector of the Minister of Housing and Local Government [C. D. Buchanan, now Professor Buchanan] said in his comment on the local Inquiry "whether the urgency of the power programme and the difficulty of finding sites for nuclear power stations are so great as to justify grevious damage to the National Park". "The issue," he said, "seems to be Park versus Power", and unlike his colleague, the Chief Engineering Inspector of the Ministry of Power, he did not feel that he could give one preference over the other. "My own judgement leans strongly to the preservation of the National Park, but questions of National Policy loom so large in the case that I think it would be presumptuous for me to make a formal recommendation".'[29] The power station is being built but the Board is taking immense trouble to reduce the adverse impact of its building on the landscape.

Another category of alien intrusions into National Parks is concerned with defence installations. In 1964 the three armed services together held over 650,000 acres of land, that is about the same size as the County of Durham. The army used over 16,000 acres in the Dartmoor National Park alone.[30] In February 1960 it was announced that a ballistic early warning station was to be erected on government-owned land at Fylingdales in the North York Moors National Park at a cost of about £43 million. The Government 'greatly regretted' the need for siting this station within a National Park, but the topographical, geographical and operational criteria were 'extremely stringent' and after detailed examination the Government was 'satisfied that there is no other suitable site in the whole country'. The National Parks Commission pointed out, at a meeting in March 1960, 'that the decision to erect this installation was plainly inconsistent with the essential purposes of the National Parks Act and expressed deep regret that the intention of the Act had been over-ridden'.[31] In a letter to *The Times* Lord Strang described the Fylingdales schemes as 'a new enormity'; and speaking of it as well as of Trawsfynydd and Milford Haven he said: 'these depredations are plainly inconsistent with the essential purpose of the National Parks Act. Her Majesty's Ministers are conscious of this. They express regret at such intrusion. They plead national necessity: we must have these things and there is nowhere else to put them. But the National Parks Commission anxiously await the day when Her Majesty's Ministers will at last firmly say, "Thus far and no farther".'[32] Protests were unavailing; Fylingdales was built and was in operation by September 1963. The claims of security are over-riding; *salus populi suprema lex*. It must be admitted again that the project has received the benefit of expert advice from landscape architects.

These are not the only blows delivered against areas of natural beauty. There are the lines of pylons which stride over the rural scene. During the summer of 1964 controversy raged over the Minister of Power's decision to run a supergrid power line on 16oft pylons (only a few feet less than the height of Nelson's Column) over the Downs in West Sussex in spite of the objections of all the local authorities concerned, and of the recommendations of the two Ministry inspectors.[33] There are the numerous masts erected on high hills by the GPO and other ministries; for instance the mast, 250yd high, erected for the BBC's sound and television services at North Hessary Tor, in the Dartmoor National Park. This was opposed by the Commission in 1953.[34] Since then these masts have appeared in many places. Miss Sylvia Crowe, the distinguished landscape architect, was reported to have said in October 1964: 'Do we really want a hatpin sticking out of every hilltop?' Again, there are the increasing demands for water to be taken from several National Parks. The most notable controversy of this kind concerned the claims of Manchester Corporation upon Ullswater at Bannisdale: a Water Bill on this subject was emphatically rejected by the House of Lords on 8 February 1962, after an eloquent speech by Lord Birkett on behalf of his native Lakeland. It was argued that what might be cheap for Manchester would not be cheap for England when the beauty of Ullswater was at stake.

But threats to the beauty and peace of the National Parks do not come only from industry, the Statutory Boards and the

Defence Services. The National Parks Commission has the duty of providing more opportunities for recreation in these areas, as well as that of preserving their natural beauty. These two aims can become incompatible, for tourists by weight of numbers can destroy the very solitude which they come to seek. The Lake District Planning Board called attention to this problem in these words: 'The danger comes from caravans and camping in the un-spoilt dales and daleheads, from the noise of motor boats on hitherto quiet lakes, from mass tourism on the fells, and from the congestion of motor traffic'.[35] A Coniston parish councillor is reported to have said: 'What are we going to do? . . . We don't want a Blackpool here, but we don't want a museum'.[36] This is a real dilemma. The increasing number of holiday-makers are bound to have an impact on the countryside including the National Parks. Yet the Parks were established partly in order to serve the needs of recreation in the open air. In *Traffic in the Lake District*, issued in September 1964 by the Friends of the Lake District it is stated that 'for the first time, tourism by sheer weight of numbers is killing what it comes to enjoy. This, in the Lake District National Park is the problem which dwarfs all others'. In the House of Commons Mr Marples said: 'I am terribly distressed about what may happen in the Lake District when the motorway [M 6] . . . links up with Manchester. A million people will be able to get to the Lake District speedily.'[37] The whole subject of holidays and recreation needs far more research than it now receives; very little is known about the precise numbers of visitors to the National Parks[38].

It is believed that over 70 per cent of British holiday-makers go to the sea for their annual holiday. For this reason alone pressure on the coast, whether in a protected area or not, has been increasing. The total length of the coastline of England and Wales is 2,742 miles: of this over 380 miles can be described as substantially built-up (1958); that means that about 13 per cent of the coastline is already developed. Between 1938 and 1958 the addition to the built-up area of the coast was 32 miles; the loss of natural coast was therefore about $1\frac{1}{2}$ miles a year. In 1968 over $4\frac{1}{2}$ million people took their holidays in coastal caravans. The demand for holiday chalets and caravan parks on the coast is insistent; these demands and the needs of commerce and industry should not be allowed to destroy the beauty and scientific interest of the coast.[39] In September 1963 the Ministry of Housing and Local Government issued a circular on 'Coastal Preservation and Development'.[40] All authorities concerned were asked to make a special study of their own coastal areas and to pay special attention to safeguarding them. Considerable parts of the coast are within the various categories of protected landscape; a large amount is included in the National Parks and Areas of Outstanding Natural Beauty. The extent of coastline protected, in varying degrees, is about 52·6 per cent of the whole, which is a substantially higher figure than the percentage of the total area of the country protected in these classes. Nevertheless the potential dangers to the coast are very great. For example planners have estimated that by 1981 about 2·8 million people will visit Devon during the summer, with a peak number at one time of 236,000. The pressure on the coast has been intensified by the demands of the Central Electricty Board for sites in the remoter parts of England on which to erect power stations.[41] There is a policy decision that nuclear power stations be sited not less than five miles from any large settlement. An important and very controversial case was that of Dungeness. In September 1958 the Central Electricity Generating Board applied to build a power station on 225 acres of the southern shore of Dungeness, immediately to the west of the lighthouse. At the Public Inquiry in December 1958 its erection was not opposed by the National Parks Commission. But eleven years earlier, in 1947, the Hobhouse Committee had suggested that Dungeness, the largest shingle promontory in England, with Romney Marsh, should become a Conservation Area. It is not, therefore, surprising that the proposal to build a nuclear power station at Dungeness was uncompromisingly opposed by the Nature Conservancy because of the 'unique and irreplaceable permanent importance to science of the land in question, for which there is no adequate substitute in Europe'. In its printed evidence the Conservancy described the area as 'the finest example of a naturally developed plant and animal community on coastal shingle in this country'. It was one of the best sites for the study of oversea insect migration, and it could be regarded as 'perhaps the most important bird observatory in Europe'. Professor C. Kidson, then an officer of the Nature Conservancy said: 'The proposed nuclear power station will, if authorised, destroy effectively the greater part of the remaining scientifically valuable area of the most significant shingle foreland in the British Isles and Europe, and one of the major coastal depositional features to be found anywhere in the world. The loss of what might appear at first sight a useless waste of shingle would be a scientific tragedy'.[42] The Council for Nature, in its evidence, emphasised that the physiographic studies at Dungeness were of practical importance to coastal engineers, and would ultimately add to our knowledge of coastal erosion so that villages on the East coast might be saved. In July 1959 the Minister of Power announced that he had sanctioned the building of the station for the usual reasons, namely, 'the growth of the demand for electricity and the great importance of implementing the nuclear power programme'. In October 1959 *Nature*[43] included a strong leader on the subject of 'National Parks and Nuclear Power Stations in Britain'. It made reference to the 'scientific tragedy' of Dungeness and concluded with the words: 'The pleas of Nature Conservancy and Council for Nature demonstrate what is involved and the price that has to be paid if any part of this small island's natural beauty, flora and fauna, or the scientific task of understanding and utilising the natural resources of Britain are to be safeguarded against sectional and transient interests'.

The Green Belts have also suffered intrusions, but it is often forgotten that the primary purpose of a green belt is to serve as a method of controlling the growth of a town. The purpose of the Green Belt is also to provide a place for recreation, but hitherto this function has been regarded as secondary. It should also be remembered that in the Metropolitan Green Belt a high proportion of land is under some form of husbandry, farming, forestry or horticulture.[44] Two controversial incursions into Green Belts have been those at Oxford and at Nottingham, but both these belts are still at the provisional stage. At Oxford the Minister allowed the erection of a spare-parts factory at Horspath, close to the existing works at Cowley but inside the proposed green belt. But he restricted the permitted area to 44 acres instead of the 70 originally requested by BMC. The plan to erect a coal-fired power station at Ratcliffe-on-Soar, within the proposed Nottingham green belt, was approved by the Minister in spite of considerable opposition. On the other hand, in June 1961, the Minister rejected the proposal to erect another coal-fired power station at Holme Pierrepoint, close to Nottingham partly because it was in the proposed Green Belt. In November 1964 it was announced that the new Minister of Housing and Local Government had given outline permission for the development of 429 acres at Hartley, near Dartford, Kent, as a new village for over 5,000 people. The site is in a proposed extension of the Metropolitan Green Belt; and part of it is marked on the County Development Plan as an 'area of great landscape value'. Kent County Council had refused planning permission on grounds of loss of amenity and agricultural land. On appeal this refusal was overruled by the Minister.

Conflicting interests concerning land use are often impossible to reconcile by compromise; yet decisions must be made by ministers. It is very difficult to value a stretch of beautiful landscape by a price in money. In a leader in *The Guardian* it was stated that 'some stretches of landscape are worth a million pounds a mile, the figure the CEGB quotes for the cost of undergrounding high-voltage power lines'.[45] It is not easy to devise a set of recognised criteria by which the preservation of amenity and 'environmental decencies' can be judged. There is no doubt that the National Parks Commission has been disappointed, if not disheartened, because considerations of beauty of landscape have been overridden in certain cases where the amenity value was admitted to be high. In the House of Lords on 1 July 1959 Lord Birkett said: 'Until it is recognized that there are places where the national interest is primarily amenity or scientific, and where power development or even defence must be secondary, neither Nature Conservancy, national parks, nor the planning of land development can have any real meaning.'

But there is a credit side to this picture. In Europe there is often generous recognition of the tremendous efforts that have been made to preserve the natural landscape in this heavily industrialised country.[46] I will give one example. Local planning authorities, by their firm control of advertising, have saved miles of road and square miles of panorama from the kind of spoilation which prevails in some countries abroad. The Gas Council and the cross-country pipe line companies have been educated into great regard for landscape factors. Mineral operators have been made aware of amenity considerations. Opencast workings of both ironstone and coal have been restored.

CONCLUSION

Over thirty years ago Vaughan Cornish advocated the formation of National Parks, where, in his own words 'the urban population, the majority of our people can recover that close touch with Nature which is needful for the spiritual welfare of a nation'.[47] By his will he left a bequest to the University of Oxford 'for the advancement of knowledge relating to the beauty of scenery'. What can be done in a university to further the ideals of Vaughan Cornish? First, by teaching we could develop a greater sensitivity to the beauty of nature. The future of National Parks and Green Belts depends on the support of an informed public opinion; for without this support they will cease to exist. It is unfortunately true that those who inhabit a district of natural beauty are often the most eager to press for its development by new industry and the building of houses. But the fact that the countryman does not always appreciate the natural beauty of the landscape which surrounds him is not surprising; he has not been educated to see it. The teachers of geography have not accomplished the task that Younghusband and Vaughan Cornish expected of them. Over one hundred years ago Wilkie Collins summed up the matter very clearly in these words: 'Admiration of those beauties of the inanimate world, which modern poetry so largely and so eloquently describes, is not, even in the best of us, one of the original instincts of our nature. As children, we none of us possess it. No uninstructed man or woman possesses it. Those whose lives are most exclusively passed amid the ever-changing wonders of sea and land are also those who are most universally insensible to every aspect of Nature not entirely associated with the human interest of their calling. Our capacity of appreciating the beauties of the earth we live on is, in truth, one of the civilised accomplishments which we all learn as an Art.'[48] Just in the same way Sir Kenneth Clark, now Lord Clark, observed that it is wrong to 'assume that the painting of landscape is a normal and enduring part of our spiritual activity'.[49] He also said that 'in times when the human spirit seems to have burned most brightly the painting of landscape for its own sake did not exist and was unthinkable'.[50] He also pointed out that in medieval times 'the average layman would not have thought it wrong to enjoy nature; he would simply have said that nature was not enjoyable. The fields meant nothing but hard work . . . the sea coast meant danger of storm and piracy. And beyond these more or less profitable parts of the earth's surface stretched an interminable area of forest and swamp'.[51] The task of awakening the people to a love of, and feeling for, the beauty of scenery must be shared; both geographers and historians, especially the historians of art, have a part to play in this. In these days the word 'amenity' is much used: it means, according to *The Shorter OED* 'the quality or state of being pleasant or agreeable'. By teaching we should endeavour to make people more conscious of the amenities of environment. It is possible to create new beauty, and Ruskin, in his inaugural lecture to the University of Oxford said that 'a nation is only worthy of the soil and the scenes that it has inherited, when, by all its acts and arts it is making them more lovely for its children'.[52] A recent President of Section E of the British Association, in his presidential address, said that 'to teach the beauties of the land for the sake of their preservation must surely be a high function of geography'.[53]

But in addition to teaching far more research is needed into all the complex problems that concern the use of land in this small island. There are many competitive claims for the use of the same land. To decide which is the best use for the community as a whole is indeed difficult, because such a great variety of considerations have to be taken into account; and in making decisions the claims of the beauty of scenery can be forgotten so easily, simply because it is impossible to give them a precise value in hard cash. The Vaughan Cornish Bequest is a useful means of furthering research into these matters, but the grants it can give are small. If a benefactor were to endow the University of Oxford with a Readership in the Geography of Land Use, the post would be of immense value, because it could extend research into the multifarious problems concerning 'the beauty of scenery as determined by nature or the arts in town or country at home or abroad', to use Vaughan Cornish's words. I want to emphasise the special need for work on our historic towns large and small as well as on the countryside. Both townscape and landscape need more effective protection, based on research. Many of our historic towns are now in great danger. During the last few years several of them have had their centres gouged out; in this process beautiful and famous buildings have been wantonly destroyed.[54]

I must also plead for more investigation of all these questions abroad as well as at home. Such problems are common to the whole of Europe and indeed to the world. But they are also local; for within thirty miles of Oxford most of the questions I have outlined can be studied in miniature. I admit there is no National Park and there is no coast. But both the Chilterns and the Cotswolds have been designated as 'areas of outstanding natural beauty'. The Berkshire Downs have long been an 'area of great landscape or historic value', but the slight protection thus given was of no avail when it was decided to drive the M 4 across the ancient cradle of England.[55] This remote and lovely belt of open country has suffered deep wounds in the last few summers of the sixties. There are Nature Reserves and National Trust Properties in the Oxford region, but there are telecommunication masts also. The City of Oxford has its provisional Green Belt and all the problems that go with it. Yet since 1965 a line of high pylons has been built across Oxford's Green Belt, and these are the gigantic monsters, normally 136ft in height, only 30ft lower than the Nelson column. They now stride across Matthew Arnold's country near Bablock Hythe, Cumnor, Appleton and the stripling Thames, dwarfing churches, villages and trees. Again since 1965 the Vale of the White Horse has been, in the words of Professor Colin Buchanan, 'despoiled by the erection at Didcot of one of the largest coal-fired power stations in Europe. This huge erection [with six cooling towers, 325ft in height and a chimney of 650ft] blots the landscape from every vantage point for many miles around'.[56] The proposed building of the third London Airport at Cublington in the Vale of Aylesbury, lying 'athwart the critically important belt of open country between London and Birmingham',[57] and only twenty-five miles north-east of Didcot, would have been, in the words of Professor Buchanan, 'an environmental disaster'. Fortunately, a surge of public opinion helped to prevent this from happening (Fig 20). In spite of this gloomy account of the Oxford region there are some items on the credit side; in recent years the Minister refused a number of applications to extract ironstone from areas in North Oxfordshire, including a working at Great Tew.

Fig 20 *Poster at Cublington, Bucks, 3 April 1971. The Government's decision not to build London's third Airport at Cublington but at Foulness was announced on 26 April 1971*

Oxford itself was once a magic city; its disenchantment proceeds apace. There is still little research on these urban and rural problems as a whole, yet they are all interlocked. And it is often forgotten how much easier it is to wound the delicate loveliness of the English lowland than that of the more rugged western regions. I will conclude this chapter with some sentences written by the late Professor G. M. Trevelyan. 'Two things', he said, 'are characteristic of our age, and more particularly of our island. The conscious appreciation of natural beauty, and the rapidity with which natural beauty is being destroyed. . . . Like the Universe, like life, natural beauty also is a mystery. But whatever it may be, whether casual in its origin, as some hold who love it well, or whether as others hold such splendour can be nothing less than the purposeful message of God—whatever its interpretation may be—natural beauty is the ultimate spiritual appeal of the Universe, of nature, or of the God of nature, to their nursling man'.[58]

NOTES

1 This chapter was delivered as a lecture, to the Oxford Preservation Trust on 13 November 1964 and was published by the Trust as a pamphlet in 1965. The lecture form has been retained but the chapter has been brought up-to-date where possible.

2 An unsigned obituary of Vaughan Cornish is in *Geogr J*, 111 (1948), 294; a biography of Vaughan Cornish by G. R. Crone is in *Dictionary of National Biography 1941–50* (1950), 179–80.

3 A complete bibliography of Vaughan Cornish's writings compiled by Andrew Goudie will appear in 'Vaughan Cornish', *Trans IBG*, 55 (1972). As Cornish wrote more than eighty papers in a period of over fifty years this list of his writings is valuable, quite apart from Goudie's interesting and acute assessment of Cornish's contributions to physical geography, especially in the study of wave forms.

4 In 1923 also Vaughan Cornish was President of Section E of the British Association and gave a presidential address on 'the

geographical position of the British Empire', which is further proof of his interest in strategic geography.

5   *Geogr J,* 56 (1920), 1–13.

6   D. W. Freshfield, in a review, states that Younghusband 'among the labours and anxieties of the march to Lhasa found refreshment in the wide landscapes and sublime solitudes and skies of the Tibetan highlands'. *Geogr J,* 58 (1921), 455.

7   *Geogr J,* 56 (1920), 3.

8   ibid, 7.

9   ibid, 8.

10 ibid, 8–9. In 1921. Sir Francis Younghusband elaborated his views in a book: *The Heart of Nature or the Quest for Natural Beauty.* This contains a report of his address to the RGS, as well as a paper on 'Natural beauty and geography' given to the Union Society of University College, London, on 17 March 1921. The book was reviewed by D. W. Freshfield in *Geogr J,* 58 (1921), 454–6.

11 ibid, 7 and Cornish, Vaughan, 'harmonies of scenery, an outline of aesthetic geography', *Geography,* 14 (1928), 275–83.

12 Cornish, Vaughan, *The Beauties of Scenery: a Geographical Survey* (1943), 15.

13 Cornish, Vaughan, 'the scenic amenity of south-eastern England,' *Trans SE Union of Scientific Societies,* 40 (1935), 11.

14 For National Parks see: Abrahams, H. M. (ed), *Britain's National Parks* (1959); Blenkinsop, Arthur, 'The national parks of England and Wales'. This long article was printed as a whole number of *Planning Outlook,* 6 (1964), 9–75, and deals with questions concerning the possible reform of National Parks legislation; Darby, H. C., 'British national parks', *Advancement of Science,* 20 (1963–4), 307–18, with a useful bibliography; *Annual Report of the National Parks Commission* from 1950, HMSO; Strzgowski, *Europa Braucht Naturparke!* (Horn, Austria, 1959).

15 For Areas of Outstanding Natural Beauty see Coppock, J. T., 'The Chilterns as an area of outstanding natural beauty', *J Town Planning Institute,* 45 (1959), 137–41.

16 For Green Belts see: *The Green Belts* (Min of Housing and Local Govt, 1962); Lovett, W. F. B., 'Leisure and land use in the Metropolitan green belt', *Journal of the London Society,* no 358 (1962), 1–16, with useful bibliography; Thomas, D., 'London's green belt: the evolution of an idea', *Geogr J,* 129 (1963), 14–24; Thomas, D., *London's Green Belt* (1970).

17 For Nature Conservancy see the annual *Report of the Nature Conservancy* from 1949, HMSO.

18 Cornish, Vaughan, 'National parks and the preservation of nature in England', paper read to the British Association at Blackpool, 14 September 1936.

19 Nicholson, E. M., 'Preservation of natural areas in Great Britain' in Jarrett, H. (ed), *Comparisons in Resource Management* (1961), 114.

20 In the *Report of the Nature Conservancy* for 1962, 31, it is stated that 'Although building or other forms of statutory development can take place [on Sites of Special Scientific Interest] only after the Conservancy have been consulted, farming and forestry operations which do not require planning permission and for which consultation is not obligatory may easily and unwittingly destroy the scientific value of a site.'

21 *The Nature Conservancy—the First Ten Years* (1959), Introduction, 1.

22 *The South-East Study. 1961–1981* (1964), 21–4.

23 Hinton, Sir C., and Holford, Sir W., *Power Product on and Transmission in the Countryside: Preserving Amenities* (CEGB, 1959), 13. See also Bracey, H. E., *Industry and the Countryside. The Impact of Industry on Amenities in the Countryside* (1963). This report was made for the Royal Society of Arts and deals with the effects of electricity generation, transmission and distribution; oil refining and distribution; and the UK Atomic Energy Authority's Establishment on the countryside; and Bracey, H. E., *People and the Countryside* (1970). A valuable and comprehensive paper by Willatts, E. C., on 'Planning and geography in the last three decades', with a full bibliography is in *Geogr J*, 137 (1971), 311–38.

24 Ninth Report on the *National Parks Commission* (1958), 72.

25 ibid, 25.

26 ibid, 75.

27 See James, J. R.; Scott, F. Sheila, and Willatts, E. C., 'Land use and the changing power industry in England and Wales', *Geogr J*, 127 (1961), 304–7, for a discussion of the siting of oil refineries in Britain.

28 *Ninth Report of the National Parks Commission* (1958), 60–1.

29 ibid, 23 and 70.

30 See *Misuse of a National Park: Military Training on Dartmoor* (Dartmoor Preservation Association (1963)).

31 *Eleventh Report of the National Parks Commission* (1960), 28.

32 *The Times*, 26 February 1960.

33 See debate in House of Lords on pylons in West Sussex, 17 November 1964.

34 *Fifth Report of the National Parks Commission* (1954), 50–3.

35 *Thirteenth Report of the National Parks Commission* (1962), 73.

36 ibid.

37 Similarly pressure on the Welsh coast increased after the Severn bridge was built.

38 This gap has been filled very successfully by Patmore, J. Allan, *Land and Leisure* (David & Charles, 1970).

39 See Dower, Michael, *Fourth Wave: The Challenge of Leisure* (Civic Trust, 1965), for the impact of the caravan on the coast.

40 Circular no 56/63.

41 See James, J. R., Scott, Sheila F., and Willatts, E. C., op cit, *Geogr J*, 127 (1961), 300–1, for a discussion on the siting of nuclear power stations; see also Crowe, Sylvia, *The Landscape of Power* (1958); and Mounfield, P. R., 'the location of nuclear power stations in the United Kingdom, *Geography*, 46 (1961), 139–55.

42 *Evidence of the Nature Conservancy for the Public Inquiry into the proposed Nuclear Power Station at Dungeness* (1958), 29.

43 *Nature*, vol 184, no 4692 (3 October 1959), 1,005–7.

44 Coppock, J. T., and Prince, H. C. (eds), *Greater London* (1964); see chap 12 (292–312) on 'the green belt' by Thomas, D.

45 *The Guardian*, 22 August 1964.

46 Strzygowski, Walter, *Europa Braucht Naturparke!* This generous tribute to British efforts is by a Professor of Economic Geography in Vienna.

47 Cornish, Vaughan, *The Scenery of England* (1932), preface 13.

48 Collins, Wilkie, *The Woman in White* (1860): p 43 in Everyman's Library edition.

49 Clark, Sir Kenneth, *Landscape into Art* (1949), xvii.

50 ibid.

51 ibid, 2.

52 Inaugural lecture delivered by John Ruskin before the University of Oxford on 8 February 1870, para 25.

53 Howarth, O. J. R., *Scott Geog Mag*, 67 (1951), 159.

54 See *Monuments Threatened or Destroyed* (1963) by the Royal Commission on Historical Monuments (England). The towns of Guildford, Poole, Gosport and York are mentioned as having suffered considerable losses but there are many others.

55 Wise, M., 'M.4: the South Wales motorway', *Geogr Mag*, 36 (1963), 435–8.

56 *Report of Commission on the Third London Airport* (1971). 'Note of Dissent', 153, para 23.

57 ibid, 149, para 1.

58 Trevelyan, G. M., *The Call and Claims of Natural Beauty*, Rickman, Godlee lecture for 1931, 8 and 30.

APPENDIX

Mackinder and Haushofer

BETWEEN THE TWO wars Mackinder's theory of the Heartland received scant attention in the English-speaking world, but it was closely studied in Germany, where it became a basic idea among the students of *Geopolitik*, of whom General Karl Haushofer was the leading figure.[1] The effect of Mackinder's work on the German school of geopolitics was described in some detail by Dr Hans W. Weigert in his *Generals and Geographers* (1941)[2] and need not be repeated here. Nevertheless it should be noted that Haushofer reproduced Mackinder's map of the 'natural seats of power' exactly as it appeared in 1904 in the *Geographical Journal*, at least four times in the periodical *Zeitschrift für Geopolitik*. Further, Haushofer repeatedly acknowledged his debt to Mackinder; in 1937 for instance, he described Mackinder's 1904 paper as 'the greatest of all geographical world views' and added that he had never 'seen anything greater than these few pages of a geopolitical masterpiece'. In 1939 he quoted Mackinder's fear lest Germany were to ally herself with Russia as an argument that the two countries should unite their powers.[3] He often repeated Ovid's maxim that it is a duty to learn from the enemy, and in his review of *Democratic Ideals and Reality* he described Mackinder as a 'hateful enemy'.[4]

It is easy to exaggerate Mackinder's influence on German politics, and in any case it must be remembered that James Fairgrieve's well known book *Geography and World Power* (1915) was also widely read in Germany. This work was translated into German by Haushofer's wife and published in 1925 with an introduction by Haushofer himself.[5] Mackinder very naturally resented any suggestion that his theories had 'helped to lay the foundation of Nazi militarism'. In a speech made on receiving the Charles P. Daly Medal, awarded by the American Geographical Society and presented to him by the American Ambassador at the Royal Geographical Society's House in 1944, Mackinder referred to these criticisms. 'It has, I am told, been rumoured', he said, 'that I inspired Haushofer, who inspired Hess, who in turn suggested to Hitler while he was dictating *Mein Kampf* certain geopolitical ideas which are said to have originated with me. Those are three links in a chain, but of the second and third I know nothing. This, however, I do know from the evidence of his own pen that whatever Haushofer adopted from me he took from an address I gave before the Royal Geographical Society just forty years ago, long before there was any question of a Nazi Party.'[7]

NOTES

1 This appendix is reproduced, by kind permission of the Royal Geographical Society, from the Introduction by the author to a pamphlet (1951) containing reprints of two of Mackinder's articles.

2 Among the numerous American books and papers on the subject of Haushofer and German geopolitics the following are especially useful: Whittlesey, Derwent, *German Strategy of World Conquest* (1942) and the same author's chapter (388–411) on Haushofer in *Makers of Modern Strategy*, Earle, E. M. (ed) (1943).

3 Outside Germany it has often been supposed that Haushofer's work helped to pave the way for the pact between Germany and the USSR in August 1939. A more recent view, and one held in Germany, is that Haushofer's political influence has been greatly exaggerated. See Crone, G. R., 'A German view of geopolitics', *Geogr J*, 111 (1948), 104–8; and Troll, C., 'Die geographische Wissenschaft in Deutchland in den Jahren 1933–45,' *Erdkunde* I (1947), 3–48.

4 *Zeitschrift für Geopolitik*, 2 (1925), 454.

5 Fairgrieve, J., *Geographie und Weltmacht* (1925), translated by Martha Haushofer.

6 Rudolf Hess, the Deputy-Leader of the Nazi Party, studied under Haushofer and also knew his son, Albrecht. See Douglas-Hamilton, James, *Motive for a Mission* (1971), for an account of the relations between Hess and the Haushofers.

7 *Geogr J*, 103 (1944), 132.

Acknowledgements

MY FIRST DUTY is to express my gratitude to my old friend Professor David J. M. Hooson of Berkeley for contributing a Preface to this book.

Next I must thank the various bodies who so readily granted me permission to reprint articles and maps that had previously appeared in their publications: to the Royal Geographical Society for Chapter Two, Chapter Four with ten maps, Chapter Five with three illustrations; and the Appendix which is derived from my introduction to an RGS pamphlet: to the Royal Scottish Geographical Society for Chapter Three with three maps: to the Geographical Association for Chapters Seven, Nine, and Ten; and for the loan of three blocks which were first used to illustrate my articles in *Geography*: to the Oxford Preservation Trust for Chapter Twelve: to the London School of Economics for Chapter Eight: to various persons listed on p 116 for Chapter Six which first appeared in a Festschrift for Professor J. H. Schultze of Berlin. Bibliographical details of the above articles are given in the first note at the end of each chapter. I acknowledge the help received from the editors of the *Geographical Journal*, the *Scottish Geographical Magazine* and *Geography*. I am also indebted to Lawrence Pollinger Ltd, to the Estate of the late Mrs Frieda Lawrence and to The Viking Press Inc of New York for permission to reprint a short extract on p 202 from James T. Boulton (ed) *Lawrence in Love: Letters from D. H. Lawrence to Louie Burrows* (University of Nottingham, 1968).

I am under a heavy obligation to my friend Dr W. H. Parker who has read the whole of the book; his comments, especially on the Mackinder chapters, have been invaluable. Professor Eila

N. J. Campbell has helped me by reading and commenting on the first two chapters. I am indebted to my former colleague F. V. Emery for reading a draft of Chapter Two and making useful suggestions.

I am grateful to Don Juan Flaquer Fábregues, Vice-President of the Ateneo of Mahón for considerable assistance during my stay in Menorca when preparing Chapter Three, and for permission to use the library of the Ateneo. Philip W. Dennis of Halesworth has generously given me considerable help both in revising this chapter on the basis of his own recent research, and in lending a photograph, Plate 5, of which he holds the copyright. In the preparation of Chapter Four on medical geography I received help and advice from the following: Dr E. Ashworth Underwood of the Wellcome Historical Medical Museum; Professor Helmut J. Jusatz, MD, of the Geomedical Research Unit of the University of Heidelberg; Professor J. S. Weiner when Reader in Physical Anthropology at Oxford; J. C. Riddell, when City Engineer and Surveyor of the City of Oxford; Dr H. C. Harley and Dr A. H. T. Robb-Smith.

When working on Chapter Five I was given much encouragement and help by the late Sir James Wordie, when Master of St John's College, Cambridge. Brinsley Ford, the great-grandson of Richard Ford, has been generous in lending photographs of his famous ancestor's drawings. Chapter Six is illustrated by a map reproduced from the author's article in *Geography* (1960) by courtesy of the Geographical Association; the map was redrawn from one made by Professor Lucien Leclaire of the University of Caen and I am grateful for his agreement. For this chapter I received useful help from J. H. Paterson of the University of St Andrews, Lewis Spolton of the Department of Education in the University College of Swansea, and Richard Brinkley.

In writing two of the chapters concerned with Mackinder (Seven and Nine) I owe much to help received from the late Sir John L. Myres. Chapter Eight, the Mackinder Centenary Lecture, was greatly enriched by the generosity of Professor Leonard M. Cantor, now of the University of Technology at Loughborough, who lent me his unpublished work, especially that on Mackinder's years at

LSE. I am grateful to many colleagues, pupils and friends of Professor A. J. Herbertson who gave me information based on their personal knowledge, which has been used in Chapter Ten. I am especially indebted to R. C. Maasz, the late Miss Nora E. MacMunn and the late Robert Aitken. The last-named collected biographical material concerning Herbertson for my address to the five hundredth meeting of the Herbertson Society at Oxford on 6 May 1958 and allowed me to use his research again in 1965. Others who knew Herbertson and gave me aid were the late Professor E. G. R. Taylor, the late Professor J. F. Unstead, Miss M. Swift, Mrs E. F. Bentley and Miss Gladys Pugh. I am indebted to T. W. Freeman for valuable comments on the first draft of that chapter and also to M. F. Robertson for useful suggestions. My thanks are also due to Professor F. Metz (Freiburg-im-Breisgau), Professor J. H. Schultze (Berlin) and Professor C. Schott (Marburg) for information with regard to Herbertson's contacts with Germany; and to Professor M. J. Wise, Dr R. P. Beckinsale, G. R. Crone and L. J. Jay.

In preparing Chapter Eleven, hitherto unpublished, on P. M. Roxby I was greatly assisted by the late Canon Charles F. H. Soulby of Liverpool Cathedral and by J. E. Allison one time Chairman of the Liverpool branch of the Geographical Association, as well as by Professor R. W. Steel of the University of Liverpool who was so kind as to make available the portrait of Roxby (Plate 13), now in the Liverpool Department of Geography. I was also helped by my friend and former colleague Professor Idris Ll. Foster of Oxford, who knew Roxby well when he was at the University of Liverpool. The last chapter Twelve on Vaughan Cornish owes much to Dr E. C. Willatts OBE, now of the Department of the Environment, who commented on the first draft; he is in no way responsible for any errors or for the opinions expressed. I am also grateful to my temporary Research Assistant, Mrs Celia Squire, who collected much material for that chapter. I am indebted to the following librarians: Dr Helen Wallis of the Map Room of the British Museum, E. J. S. Parsons, now Secretary of the Bodleian, R. F. Tandy of the Bodleian's Map Room, and Miss Elspeth Buxton of the Oxford School of Geography. I received much help from Miss Mary Potter (cartographer) and P. C. Masters (photographer) both of the Oxford School of Geography. Others who have given me assistance include Professor Jocelyn M. C. Toynbee, Professor Arthur Davies, John Patten, Andrew Goudie, Harold Loukes and Professor John Warkentin of York University, Canada. General Sir James Marshall-Cornwall kindly read and commented on the Introduction. Professor Carl O Sauer, and Dr David Stoddart generously allowed me to quote from their letters. I am also grateful to Sir John Betjeman for some helpful suggestions.

The copyright of the owners of pictures and photographs is acknowledged in the list of illustrations on pp 9–11. Mrs Peter Tabor generously allowed her Richmond portrait of H. W. Acland (Plate 10) to be photographed and I am grateful to my cousin, Geoffrey Gilbert, for doing this. My thanks are due to F. R. Maddison, Curator of the Oxford Museum of the History of Science, in connection with Plate 2; to Dr S. P. W. Chave of the London School of Hygiene and Tropical Medicine, for providing the portrait of John Snow (Plate 9); and to Sifton Praed & Co Ltd for the portrait of Vaughan Cornish (Plate 16) taken from Cornish, Vaughan, *Kestell, Clapp and Cornish. Records of Home Life and Travel* (1947).

My friend and former pupil, J. Allan Patmore, helped me to plan this book and then gave me constant aid and encouragement to complete it. He generously provided the drawing of Conservation land (Fig 19), partly derived from his Fig 62 (p 194) in *Land and Leisure* (David & Charles, 1970); and also helped me obtain the Liverpool photographs (Plates 13 and 14 (*b*)).

It has been a very pleasant experience to work with Dr Brian Harley, of David & Charles, in the preparation of this book; nobody could have been more patient and more helpful with scholarly advice.

EDMUND W. GILBERT

Old Cottage, Appleton, Berkshire
May 1971

Index

Bottomley, Lady Alice, 154
Boyce, Edmund, 101
Bramshill Park, 44
Brendan, St, 26
Brigham, A. P., 193
Brighton, 146; three climates of, 81; health map of, 98
Brinkley, Richard, 126n
Bristol, 220
British Association, Birmingham (1866), 143-4
Broad Street, London, 81-2
Brontë, Charlotte, 119
Brontë, Emily, 119
Bromsgrove School, 211, 212
Bruce, W. S., 186
Buchan, Dr Alexander, 186
Buchanan, C. D., 26, 28n, 29n, 243, 252
Buckden, 211, 218
Bulkington, ('Raveloe'), 121
Burford, 45
Burslem, 122
Bygott, John, 192
Byng, Admiral, 72

Cader Idris, height of, 55
Camberley, 232
Cambridge, City of, 160; Green Belt, 236
Cambridge, University of, Geography at, 31, 132, 134, 135
Camlachie, 148
Campbell, Colin, 64, 67
Canals and Waterways, Royal Commission on, 197
Canterbury, 146
Cantor, L. M., 140, 162n
Capel, Harriet, 106
Carfax, 93, 95
Carnarvon, Earl of, 112, 115n
Carpenter, Nathanael, 48
Caskgate Street, Gainsborough, 162n; *illustration*, 156
Caswell, John, 22, 50, 54; measurement of heights by, 55, 58n
Central Electricity Generating Board, 246
Charles, Prince of Wales (later Charles I), 46
Chaudhuri, Nirod C., 116, 126n
Childs, W. M., 146
Chilterns, 251

China, 215; Roxby's publications on, 217-19
Chinnor, 191, 200
Chisholm, G. G., 149, 150, 176
Cholera, in Great Britain, 81; in Iceland, 81
Chorography, definition of, 53
Christ Church, Oxford, 32, 35, 36, 141, 142, 145, 146, 213-14
'Christminster', (Oxford), 120
Chrystal, G. O., 182
Ciudadela, 59, 65, 67, 71, 73, 74; *illustration*, 52
Ciudad Rodrigo, 106; *illustration*, 104
Clark, Sir James, 80-1
Clark, Sir Kenneth, (Lord Clark), 19, 250
Cleghorn, George, 64, 71
Clothworkers' Company, 37, 38
Coastline of England and Wales, 245-6
'Coefficients', the, 167
Colby, Col T. F., 94
Collins, Wilkie, 249
Comberton, 56
Cook, Capt James, 19, 22, 140, 141, 161
Copeland, Ralph, 183
Copernicus, 54
Cornish, Vaughan, 20, 25, 26, 31, 249; Life, 228-32; bibliography of his writings, 253n; *portrait*, 208; visits Gainsborough, 162n
Cornish, Vaughan, Bequest, 227, 251
Cornwall, 80
Cosmographie, 46, 47-50
Cotswolds, 251
Countryside Act (1968), 233
Countryside Commission, 233
Cowley, 248
Crabbe, George, 24, 213
Craster, H. H. E., 214
Crawford, O. G. S., 192, 199, 200, 202
Crillon, Duc de, 73
Crone, G. R., 205n, 232, 253n, 258n
Crowe, Sylvia, 244
Cublington, airport, 26, 252
Cumnor, 251
Curzon, Lord, 141, 163n, 216; and Oxford School of Geography, 200-1; on *Frontiers*, 200-1
Czajka, Willi, 195

Darby, H. C., 121, 126n
Darbishire, B. V., 172, 185